D1665506

Transacting Sites of the Liminal Bodily Spaces

Transacting Sites of the Liminal Bodily Spaces

By

Catalina Florina Florescu

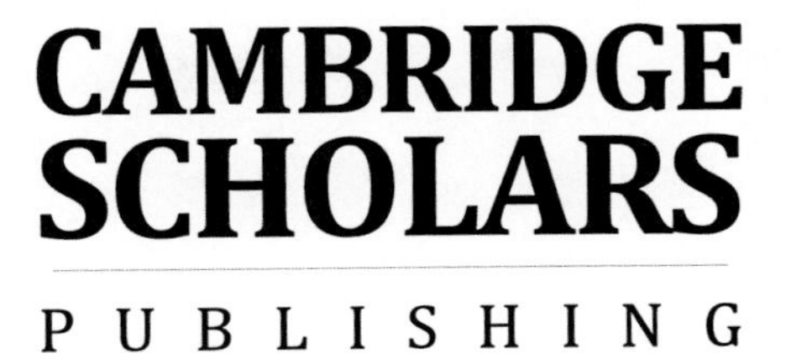

Transacting Sites of the Liminal Bodily Spaces,
by Catalina Florina Florescu

This book first published 2011

Cambridge Scholars Publishing

12 Back Chapman Street, Newcastle upon Tyne, NE6 2XX, UK

British Library Cataloguing in Publication Data
A catalogue record for this book is available from the British Library

Copyright © 2011 by Catalina Florina Florescu

All rights for this book reserved. No part of this book may be reproduced, stored in a retrieval system, or transmitted, in any form or by any means, electronic, mechanical, photocopying, recording or otherwise, without the prior permission of the copyright owner.

ISBN (10): 1-4438-2693-6, ISBN (13): 978-1-4438-2693-8

TABLE OF CONTENTS

ACKNOWLEDGEMENTS

I would like to thank my mentors, Professors Thomas Adler and Elizabeth K. Mix, for their constant help, support and words of encouragement and improvement; Professor Keith Dickson, whose rigorous self-discipline techniques I have tried to emulate and apply in my own life; Professor Floyd Merrell, for suggesting me a series of books that bloomed my senses; my husband, Professor Ionut Florescu, for teaching me the complex, contradictory nature of being "absent-present"; *my mother-father-grandmother who guided me spiritually and would be proud of me, if they were still alive*; Mari, my sister, whose former passion for writing I continue; all the artists used in this study whose works have been a lesson in endurance for me, too; Mimi, my one and only alter ego; and, finally, Mircea Gruia—my son and **love**—whose cries, laughs, joys and frustrations, first teeth, steps, and words have literally accompanied me throughout the challenging and exhausting process of assembling and then writing this book.

I also *thank*

"Ciulendra" (a very rapid and passionate traditional Romanian song),
k.d. lang's "Constant Craving"
Goran Bregovic,
Satie,
Keith Jarrett,
Yo-Yo Ma,
Frank Sinatra's "My Way,"
Gheorghe Zamfir,
Alicia Keys,
Johann Strauss's "Radetzky March,"
Edit Piaf,
Nina Simone,
Maurice Ravel's "Pavane pour une infante défunte,"
Norah Jones,
Giacomo Puccini's aria "Nessun Dorma,"
Diana Krall,
David Bowie,
B.B. King,

Leonard Cohen,
Tōru Takemitsu,
Annie Lennox,
Queen's "The Show Must Go on" and "Bohemian Rhapsody,"
Aretha Franklin,
George Enescu's "Rapsodia Română,"
John Lee Hooker,
and so many other musical pieces that have kept me company when in *silent* pain, or, at least, in doubt about my identity.

INTRODUCTION

We are a bundle of chemical reactions, *if anything*; some are still active, others retroactive; some are reversible, others irreversible; some have quick results, while others cause long-lasting effects; some are productive, while others fail to produce anything. Therefore, when we move not only do neuronal assemblies tickle (within) our brains, but also our unlimited scents change their intensity while interacting with the world and other creatures. How are our bodies structured? In *Phenomenology of Perception* (2002), Maurice Merleau-Ponty provides one possible answer by arguing that "The outline of my body is a frontier which ordinary spatial relations do not cross. This is because its parts are inter-related in a peculiar way: they are not spread out side by side, but enveloped in each other" (112). As the title of my book, *Transacting Sites of the Liminal Bodily Spaces*, suggests this work is about enveloped and enveloping spaces, about zones of easy and not so easy accessibility within our bodies, about what we know about our bodies, as well as what we thought we knew about them.

Have we ever trusted our bodies, or have we always tried to control them? Part of my interest in writing this book is directed toward the body dying; the body opened in surgery, or through MIRs, CATs, and sometimes in autopsies; the body preserved through computerized images such as those created by the Visual Human Project; the metonymic body that continues to live in another body through organ replacement; bodily parts cast in silver, and then abandoned in a museum. Arguably, we perform our lives transacting and negotiating our needs with the needs of other people. But does transaction cease once the body is dead, or close to dying?

I must have heard and read the word "transacting" several times in my life until it has finally started to acquire a particular importance for my work. The very first lines of Shannon Sullivan's book *Living across and through Skins: Transactional Bodies, Pragmatism, and Feminism* (2001) read as follows:

> Thinking of bodies as transactional means thinking of bodies and their environments in a permeable, dynamic relationship in which culture does not just effect bodies, but bodies also effect culture. [...] In [John]

> Dewey's words, '[t]he epidermis is only [...] an indication of where an organism ends and its environment begins. (3)

Later, she writes that "When Dewey speaks explicitly of bodies, he prefers to talk of 'body-mind' [where] both terms designate bodies as transactional. In contrast to much of traditional philosophy, Dewey describes mind as continuous with and emerging out of physicality" (24). While physicality is of a paramount importance for my work, for the moment let me offer a first stop or cut—if you will—by explaining what I mean by "transacting" and how I envision using it throughout my book.

As public persons, we act as well as transact our own and others' spaces. As actors and actants, we perform several roles, just as several roles per-form us. We embody certain functions (social, political, cultural and intimate), and we are equally acted upon by these functions. Related to the image discussed earlier, we could be the pebbles thrown into the water, but we could also be the water where ripples are formed. As children, we learn a language through whose morphology and syntax we try to express what we feel and think; as adults, we discover that sometimes the language we have acquired is not adequately designed to express our thoughts and feelings. Instead, we rely on touch, gestures, and other non-communicative means to connect with ourselves and others. Consequently, for me "transacting" signifies more trans-acting; the hyphenated version of the word alludes to the coexistence of multiple zones of hiddenness and unhiddenness within and around us. According to Merleau-Ponty,

> In so far as I inhabit a 'physical world,' in which consistent 'stimuli' and typical situations recur, my life is made up of rhythms which have not their *reason* in what I have chosen to be, but their *condition* in the humdrum setting which is mine. Thus there appears round our personal existence a margin of *almost* impersonal existence, [...] which I rely on to keep me alive; round the human world which each of us has made for himself is a world in general terms to which one must first of all belong in order to be able to enclose oneself. (96-7)

If I read Merleau-Ponty's usage of the verb "to belong" correctly, then it means that we form and develop our identity when we are active *part*-icipants in a larger community. However, he also mentions those moments beyond our control and/or power that generate around our *rim* of existence an impersonal sensation.

But what if, for some reason, we cannot participate in life? Does this automatically imply that we are excluded from society? Sue E. Cataldi argues that "The subject who sees is a being who moves. Only to such a being will space appear as articulated into regions of removedness" (43).

Could we infer that in some unfortunate cases (e.g., the convalescent body, the severely drugged body, and the paralyzed body—just to name some of which I use as examples in my work), these persons' sense of space is removed from their perception of being? Could we further infer that this is how they commence to be disinterested in their belonging-to-the-world? Finally, could we also suggest that their sense of space becomes exclusively internalized? While attempting to answer these questions is highly provocative, if not problematic, nonetheless pursuing answers allows me to introduce three concepts fundamental to my book: absence, interiority, and thresholds—all three being arranged around the main theme of bodily liminality.

However, to understand these concepts (and how I apply them), first it is useful to address this question: Where and how could we define the space(s) of our bodies? We all need and have our space. We populate this space with many objects; we welcome some people in, while we leave others out. But most importantly of all, within us there lies the vast, more often than not ignored space of our liv*ed* lives; the space of the body that has become a story mixing identities, misplacements, and partial recoveries. Alongside the space of the lived moments of our lives, there is the space of incubating desires, just as there are spaces never to be explored. As Bryan Turner points out, "Because embodiment has in fact many dimensions, one can talk about having a body in which the body has the characteristics of a thing, being a body in which we are subjectively engaged with our body as a project, and doing a body in the sense of producing a body through time" (281).

The body as a project has an unfinished quality, a suspension of disbelief *vis-à-vis* its mortality, and it also reminds me of the Deleuzian and Guattarian "BwO." In *Medicine as Culture: Illness, Disease and the Body in Western Societies* (2004), Deborah Lupton notes, "The body-without-organs is much more than the material or anatomical body (this is, the body-*without*-organs). It is the body self, the 'self-inside-the-body,' a phenomenon that is cognitive, subconscious and emotional as well as experienced through the flesh" (24). However, when we live with a body whose inside begins to collapse and deteriorate, the body becomes very compartmentalized *with* organs. In other words, we could easily engage in sophisticated, almost endless discussions about the unfinished quality of our bodies; in reality probably nothing unsettles and troubles us as much as when our bodies fall (seriously) ill; those are the moments when we wish our bodies were a "finished" project, namely, remained the same, unaltered.

Arguably, the issue of liminality is not novel. Thresholds, doors, windows have been considered dual objects for centuries. These objects offer both an entering and a closing, a shared space that because of its ambivalent nature could function either way at any moment. Moreover, these ambivalent, Möebius-like spaces have the quality of revealing and/or withholding other spaces, playfully engaging themselves and us in an in and out movement. Furthermore, "An occluding edge has a paradoxical status: 'It both separates and connects the hidden and unhidden surface, both divides, and unites them'" (Cataldi 32).

Could we read our bodies as liminal? Initially, I thought liminal bodies were only those confronted with and considerably altered by a serious illness (e.g., cancer, AIDS, Alzheimer's). Then, the more I read and reflected on this matter, the more I realized there is an intriguing, indeterminate quality of the liminal bodies. Consequently, I have enlarged my definition of liminal bodies to liminal bodily *spaces* transacting not only their own territories but also other(s') territories, due to a.) the uncertainty in the etiology of AIDS and cancer, as well as the inconsistency in their prescribed treatments; b.) the oscillating debates over what constitutes health and what not; c.) paralyzed bodies confined to a limited existence via machines; d.) bioengineering techniques that seek to reform the body via prosthetic devices, artificial organs and tissues; e.) organ transplants and the now very difficult matter of clearly defining death; and f.) the computer, virtual reality and the birth of the posthuman.

I have found it useful to enhance the issue of bodily liminality through what may seem at first contradictory, namely through habits (e.g., how we move our bodies, what types of expectations we have from our bodies, etc.; in other words, all those activities through which our bodies' reactions have become predictable). Sullivan notes that "The familiar, deeply engrained nature of habits such as locomotion and language often only becomes something of which one is conscious when they are interrupted" (31) and "Habits are formed in and through an organism's transaction with its environments" (33). On the other hand, I argue that no matter how deeply ingrained our habits are, the encounter with an illness causes the unfamiliarity of our own bodies to surface. As we shall see, the issue of bodily liminality pushes to the extreme the notion of habits, and it challenges our understanding of habitual experience. For the moment, however, it is sufficient to remark that no matter how well we think we have managed to transact the meanings and needs of our lives, when in pain we start to learn a new vocabulary; the body has a new syntax and morphology, while the mind still dwells in the space of its habits. There is an indisputable, and in most cases, quite inconsolable tension between the mind and the body in pain.

Just as it is difficult for the mind in pain to synchronize with the body in pain, the convalescent body's identity is also problematic to redress. In *The Absent Body* (1990), Drew Leder notes that the gaze of the other thematizes our bodies. Since my book focuses primarily on bodily spaces seriously affected by illnesses, I propose to move the issue of thematization a step further by suggesting that illness itself thematizes a body.

Practically in all the works that I propose for analysis, the characters affected by a serious illness--those who have become significantly altered/Other--speak about their disappearance from the social space. In the past, when someone was diagnosed with cancer or AIDS, it was thought that the whole society was contaminated with cancer or AIDS, and thus society graciously partook in the pain of the other. Needless to say, those were not lasting discourses simply because they were based on empty analogies and false hopes. Arthur Frank, who had cancer, proposes that we think of society as being in remission; in doing so, we may more accurately parallel the phenomenon of cancer, which is frequently in remission. Even more importantly, we should take care of our bodies so that we do not upset the social equilibrium: "[i]f people choose to ignore health risks they are placing themselves in danger of illnesses, disability and disease, which removes them from a useful role in society and incurs costs upon the public purse" (Lupton, *The Imperative* 90).

While conducting my research, I have also noticed that discourses on health have stressed the importance of the return to an initial bodily and mental condition for those whose bodies have been afflicted with a potentially life-threatening illness. It is not only that the onset of an illness thematizes a body, but also a return to a previous equilibrium is desperately wanted and sought. Hans-Georg Gadamer states that "The sick person is no longer simply identical with the person he or she was before. For the sick individual who 'falls out' of things, has already fallen out of their normal place in life. But the individual who now lacks and misses something previously enjoyed still remains oriented towards returning to that former life" (42). The movement of the seriously altered bodies toward a return to their previous conditions and privileges is not only not synchronous with their lives (which invariably move forward), but also a misplacement of their bodies (how could they physically go back?), as well as a misapprehension of the significance of their illnesses (and the possible ways to understand and cope with them). However, Gamader is accurate when he writes that "The life of the body always seems to be something which is experienced as a constant movement between the loss of the equilibrium and the search for a new point of stability" (78).

However, since I have never experienced any major discomfort, any considerable alteration to my physicality, my book is only a "travelogue" through the pain of others. Richard M. Zaner, a physician, in *Conversations on the Edge: Narratives of Ethics and Illness* (2004) remarks the importance of interacting with those who are so ill they are confined to bed. He argues that the relationship between patients and doctors, as well as between patients and their relatives, is "[d]eeply *reflexive*. Kierkegaard was right: the reflexive is at the heart of being a self. […] Seeing a [patient] is seeing [the patient] seeing me" (35). The question is whether or not we could function as choral figures in the other people's drama. How deep and/or steep is the edge when we conduct these conversations? Even more importantly, how many of us have had the chance to actually sit next to someone who lies in bed, and whose mind and body experience an advanced state of deterioration? Crouching our bodies to occupy as little a space as possible, we change our posture; our minds are inundated by a flood of sensations, which do not always concern the patient him/herself. Then, how do we transact with the pain of these persons/patients? According to Merleau-Ponty,

> I perceive the other as a piece of behavior, for example, I perceive the grief or the anger of the other in his conduct, in his face or his hands, without recourse to any 'inner' experience of suffering or anger, and because grief and anger are variations of belonging to the world, undivided between the body and consciousness, and equally applicable to the other's conduct, visible in the phenomenal body, as in my own conduct as it is presented to me. But then, the behavior of another, and even his words, are not that other. The grief and anger of another have never quite the same significance for him as they are for me. For him these situations are lived through, for me they are displayed. (414-15)

Could doctors perceive how to more effectively decode our behavioral language and decipher our unspoken needs? Even more importantly, they should not be the only ones involved in the process of healing. Few persons/patients actually consult a therapist along with a physician, even though the physical wound/anomaly/discomfort should be treated at the emotional level, too. Lupton notes that centuries ago, "Along with the physician, traditionally considered a public health expert, other specialists were needed to collaborate in the goals of the social hygiene movement, including pharmacist-chemist, veterinarians, engineers, architects, scientists and administrators. […] It was not until public health became professionalized that medical practitioners began to play a more dominant role" (*The*

Imperative 25). Because doctors are still invested with this power to heal, when we fall ill we almost exclusively ask them to cure us. Entering their office is one decisive moment when we finally refer to ourselves as being aware of our embodiment.

However, Katherine Young remarks that "In fact, storytelling is exceedingly rare on any medical occasion. [...] Patients produce 'replays' that do not achieve the status of fully-fledged narratives, stories that do not come to an end... [...] In replays, there are no beginnings and ends, only ongoing events. So events do not appear to take a direction" (69-70). Undeniably, there is more than diagnosing someone and prescribing a treatment. In the doctor-patient encounter, the former has the power not only because institutionally s/he is placed in a more advantageous position, but also because we—whether ill or healthy—do not know the inflections of our bodies. Accordingly, we are disoriented in a doctor's office, and we feel we have just flunked a very important exam; we discover that most of us are practically illiterate when it comes to our bodies and their signs, symptoms, and functions. Those are the moments when we realize that while still in the body, we feel somewhat out, exiled, pushed away from our comfortable, habituated lives.

All the persons analyzed in my work speak of this bodily exile. They reveal to us the space of their new body; please note, however, that this is a new *and* deteriorated space unlike the usual commercials' "new and improved." Susan Sontag has noticed that the iconography of pain is not new:

> The sufferings most often deemed worthy of representation are those understood to be the product of wrath, divine or human. (Suffering from natural causes, such as illness or childbirth, is scantily represented in the history of art; that caused by accident, virtually not at all—as if there were no such thing as suffering by inadvertence or misadventure). (*Regarding* 40)

On the other hand, she also remarks that today "[t]here is a mounting level of acceptable violence and sadism in mass culture: films, televisions, comics, computer games. Imagery that would have had an audience cringing and recoiling in disgust forty years ago is watched without so much as a blink" (*Regarding* 100-101).

Almost exclusively, the persons selected for my book have bodies so deteriorated by illness that they transact (or should I say trade?) life over death. Through the tedious course of their illnesses and the various treatments, these individuals eventually become liminal in the extreme understanding of the word. Without being tactless, it is a very confusing task to interpret them as having bodies in the traditional rendition of the word, as it may be more accurate to interpret them as bodily spaces reduced by illness to a more and more minimal space. Thus, whether they speak their pain as fictional characters or as real persons in memoirs, whether we partially acknowledge their pain in photographs or in films/documentaries, their altered bodily spaces as well as their potential disappearance are so significant, that they unintentionally open a useful debate over the limitations of the healthy body. What they suggest is that the demarcation line between being healthy versus ill is precariously drawn; that, whether or not we like to accept it, our bodies' matter is eventually breakable, despite our efforts to maintain it whole.

I also direct my attention on bodily spaces because recent medical discourses have suggested that just as one body undergoes several developments and embodiments throughout its existence, so it seems that the body does not experience one death, but deaths. More explicitly, not only do autopsies confirm the accuracy (or not) of a clinical diagnosis and its treatments, but they also describe the subsequent malfunctions of the bodily tissues (malfunctions of which a patient might or might have not been aware).

Therefore, the human body is a collection of bodily spaces, of which none is truly opaque or, unfortunately for us, unbreakable. Moreover, some of these spaces mature and evolve purposefully; others grow enervated and/or get sick. However, our bodily spaces' sometimes independent development from our desires and strenuous self-discipline exercises could explicably become frightening sources of anxiety for us. (Throughout my research I have encountered countless variations of the following question: I can control so many aspects of my life, why cannot I fully control my body?) Furthermore, the body to be comprised of bodily spaces understood as thresholds. Consequently, the thresholds are not those spaces that exclusively delineate the outside from the inside anymore; they could also be those spaces situated within other spaces which we do not see, but which sometimes surface themselves as signs *of something*. The body finally reveals its interiority.

Tracing the split between mind versus body in her book *The Flight to Objectivity: Essays on Cartesianism and Culture* (1997), Susan Bordo writes:

> The idea that Descartes discovers in the second Meditation is the *cogito*, which assures Descartes that there is one judgment about the nature of *things* that insures its own correspondence with external reality—one, indeed, for which the issue of correspondence between idea and reality cannot arise. That judgment is *sum*. The indubitability of events in

> consciousness now plays a role in establishing the special indubitability of *sum*: whatever I am thinking, I cannot doubt that I think. That 'I think' entails '*sum*' cannot be doubted [...] It is generally believed that the search for indubitability ends here, with the discovery of the *cogito*. (25)

Let me insert my reading in regard to René Descartes' *Dubito, ergo cogito; cogito, ergo sum* (or "I doubt, therefore I think; I think, therefore I am"). Does *dubito* truly leave us once we start thinking? The quintessential discoveries of our interiority and the invention of the mind have kept as their "raw" element the introduction of doubt. Particularly related to my book, many a time I have recorded these questions/doubts: Was that diagnosis accurate? Would the prescribed treatment prove efficient? Would I be able to recover? Once doubt has managed to sneak into our *mens cogitans* (Latin for "the thinking mind"), most likely it remains there. In other words, it is very difficult for us to remove doubt. Then, probably the invention of the mind is a daily practice in which we constantly alter the *sum* of our doubts. However, how comforting is it for us to know we are endowed with a thinking mind when we face the conflict at the level of our sensory data? Do we rely on *dubito* or *cogito* when we realize we cannot explain everything that we experience, and for that matter that science cannot either? This degree of uncertainty that weight upon us becomes even more burdensome for those persons/patients whose lives may acquire a sudden new meaning depending on blood tests, CAT scans, and/or iatrogenic complications.

Therefore, the body is in its physical doubt just as much as our mind is. Through the experiences of the persons whom I examine in my work, I have been able to make some major personal discoveries. For example, now I can say *without* doubt that there was a time in my life when I naively believed that for every cause there was a corresponding effect. Then I grew out of this phase, and realized that for every cause there could be effects, side-effects with side-affects, omissions of effects. There was another time when adverbs meant exactly what they were supposed to mean, thus never exceeding the limited space of their linguistic dimension. For example, "always" was *always* "always" and nothing more, since in my mind no other word was supposed to share in its meaning. There was still another time when certain concepts were caught in one unshakable *imago* (e.g., "the heart," "the house," "the tree"). There was yet another time when I could not make sense of anything unless I relied on its verbatim definition.

However, as I have come to realize lately, to live is to experience our lives *as passing*, to try many things, to fail, and to ach(iev)e. Then, our desires fluctuate not only from the corporeal to the emotional site, but also from past to present into a new reading of the present. As Elizabeth Grosz points out, "The future is the domain of what endures. But what endures, what exists in time, and has time as part of its being is not what remains the same over time. The past is what endures, not in itself, but what is open to becoming, to something other" (*Histories* 16).

This book is divided into six chapters and it introduces three major concepts: person/patient, *fleshbacks* and *fleshforwards*, and *le corps déjà-vu*. The first concept addresses the inseparable reconfigured identity of someone who has been diagnosed with an illness and has to cope with it physically and emotionally. More specifically, in order to avoid the technicality of the expression "medical case," I propose to look at patients as persons/patients transacting the newly added dimension to their physical and social identity (a more elaborate perspective on this phenomenon is provided in chapter two of this book). The second term addresses our desire for control as well as immortality. We do not want to live our lives on the edges of our bodily sensations as remembered or amassed via memories and repressed desires; instead, we want to live instantaneously with our emotions and events, just as computers inhabit and dominate the present moment. Put differently, we are tired of living at the mercy of the inconsistency of our flashbacks. We want to somehow create *fleshbacks* and especially *fleshforwards*, folding and unfolding, and above all lasting (a detailed view of these two concepts appears in the first chapter). Finally, *le corps déjà-vu* brings into discussion contemporary approaches to perceiving and cataloguing the human body, its borderline identity (posthuman and hybrid) and its increased tendency of becoming something overpowered and controlled by mass-media.

My conclusion is an invitation to analyze the a(n)esthetics of the healthy body, and the limits of an open culture and its institutions. What is health? To find one satisfactory definition for health is a vexing matter. Gadamer was cautious when he entitled one of his books *The Enigma of Health* because he realized he could not say with certainty what health is. I cannot say that I have found a precise definition either. However, it would be beneficial for us if we realize that we have been focusing too much on the fluctuations of our minds, disregarding the needs of our body, or equating our corporeal needs with our mental ones. I am also interested in analyzing the discourses of the contemporary body commissioned by the vast industry of mass-media. This type of body has started to direct itself toward frugal pleasures, of which eating is probably by far its most common "enemy"; we also work harder and do not have much time to relax.

Consequently—unlike those seriously affected by illnesses—a body constantly interrupted by fear eventually runs on empty, becomes a *corps-déjà-vu*, and thus moves toward different types of minimal and liminal topology. Therefore, from eugenics through the human genome project to contemporary ads on how we (should) maintain our healthy status, it has been argued that the future of health is highly dependent on people's self-control and self-surveillance. As each chapter progresses, I reflect upon how this type of control is not only a reversal of the Panopticon, but also a way of exposing our bodies to a constant fear, controlling and thereby not indulging the body in some of its minor, yet necessary, needs. Put differently, this type of fear (e.g., the fear of getting sick) may become a perverse means to punish our bodies.

CHAPTER ONE

LE CORPS PERDU/ LE CORPS CONTINUÉ

> [p]erhaps the entire evolution of the spirit is a question of the body; it is the history of development of a higher body [...] The organic is rising to yet higher levels. (Nietzsche 358)

> In the end there is no end: totality is not achievable. (Cubitt 365)

> [In oral cultures,] [t]he body is not yet a mechanical object, but a magical entity, the mind's own
> sensuous aspect. (Abram 15)

> If I were told: By evening you will die, so what will you do until then? I would look at my wristwatch [...] I would read a chapter in Dante [...] and see how my life goes from me to the others, but I wouldn't ask who. (Darwish 119)

If there were a magical potion that could secure our immortality, would we take it? If there were a method, exercise, or technique that could keep intact (unaltered) our past experience while not interrupting our development, would we use it? A long time ago, the alchemists ardently believed that the elixir vitae could "prolong human life indefinitely" (Berman 77). In scientific times, such as ours, we refuse to think that any elixir could postpone our death or secure our immortality; and yet, we desperately seek an alternative. Robert T. Eberwein contends that our dreams have a phasic nature and that through them we step into "the D (desynchronized) state" of our minds (17). Dreams, as well as profound reveries, disclose the unfathomable structure of our sentient dimensions. Many of the arguments of this chapter revolve around the "D state," where the mind and body are not in binary opposition, but ex-tend from what we usually take to be their delineated contours. According to Morris Berman, "[t]he subject/object distinction of modern science, the mind/body dichotomy of Descartes, and the conscious/unconscious distinction of Freud, are all aspects of the same paradigm; they all involve an attempt to know what cannot, in principle, be known" (148). Here I am less interested in the cognitive, often misleading nature of our Cartesian reasoning, but in how we could dive experimentally into the deepest waters of our psyche. Like dreams and desires—whose linearity, logic and coherence are rarely contested—sketching in imagination an immortal, immaterial body is attainable.

Ever since Sigmund Freud's psychoanalytic theories, we have interpreted dreams as the loyal/royal paths to our unconscious; but in this chapter of this study I am interested in those dreams, fantasies and revelations that we undergo with eyes wide open. We dream of what is absent, more specifically of a form of the body/mind that is not yet created, but whose design is conceived as immortal. However, today we are too immersed in our daily activities, duties, deadlines and commitments to listen to our needs. In *The Enchantment of the Modern Life: Attachments, Crossings, and Ethics* (2001), Jane Bennett affirms that science is the main culprit for this state of disenchantment. She cites Max Weber's essay "Science as Vocation" (1918-1919), where he argued that "'In science, each of us knows that what he has accomplished will be antiquated in ten, twenty years.' […] As this logic of the perpetually receding goal is generalized throughout culture, […] time itself comes to be figured as a moving line, a vector approaching but never reaching a vanishing point" (60). Theories about the frequent changes experienced by science are also elaborated by Martin Heidegger in *The Question Concerning Technology and Other Essays* (1977), where he suggested that we are situated between two axes of preservation and enhancement which "[m]ark the fundamental tendencies of life […]. To the essence of life belongs the will to grow, enhancement. Every instance of life-preservation stands at the service of life-enhancement. […] Anything that is alive is therefore something that is bound together by the two fundamental tendencies" (73).

This chapter is designed as a triptych of the desire for immortality. The three artists' works analyzed here— Natalie Horne's portable sculptures and excriptive-like images of the body, Félix Gonzales-Torres's perishable sculptures, and Pedro Almodóvar's film *All about My Mother* (1999)—constitute an answer and a provocation to the recurrent dream of immortality. Inspired by the experimental quality of their works, I develop the theme of "parallel-being," namely that being whose contours and contents are dreamed of, desired, and finally become imaginatively palpable and attainable. We ex-tend ourselves through words and in images, in persons and memories—while our arteries create their interior symphony—until we develop a parallel version of ourselves.

More explicitly, in times like ours, when we constantly negate the meaning of "real-ness," it seems only natural to create such a parallel version, which speaks about our obsession of penetrating the other side of the mirror, thus entering into Wonderland. There, like Alice, we could sojourn indefinitely via our imagination because time has lost its rigidly-mounted-on-its-screen twenty-four hours. Time has discovered its seconds transformed into sheer sensation, a kind of mind's elixir to the body's constantly decaying form. Time has started to dissolve one's body's (desired) pleasures and deposit them into the mind's memories. Each of the three artists' works analyzed here prepares the body for its demise, distinctively securing its immortality and thus reflecting that "When it comes to the post-mortem self, a more coherent body of evidence suggests that individuals can remain *present* within social life after their death" (Hallan 63).

Flesh-backs and *Flesh-forwards* of the *Body Redivivus*

Ego spem pretio non emo
(Terence)

Figure 1. Natalie Horne. *Vital Necessities* (2005).

Before the electronic delirium, we used to hold books in our hands and read them; our ears were not caught in a droning white noise, our hands did not touch a keyboard connected to a screen with shifting images, and our eyes followed the linearity of the old-fashionedly aligned words on a book page. In anatomical atlases, however, we still examine the impersonal, standard image and structure of our bodies. When advanced techniques have started to be employed in medicine—X-rays, MRIs and

CATs—the flesh began to lose its mythical opacity. Its interior was finally illuminated. In her book, *The Visible Human Project: Informatic Bodies and Posthuman Medicine* (2002), Catherine Waldby remarks, "The anatomical body is co-emergent with the anatomical text, and only begins to take shape when the organism is itself addressed during the dissection as a form of book—as a readable terrain whose telling surfaces can be transcribed cartographically, laid out serially in the serial space of the book" (63). This section analyzes the body as a collection of intimate, yet portable and readable objects, by centering its critique on the aesthetics of the body *redivivus* in conjunction with Natalie Horne's 2005 exhibition *Immortalis: Inside Green, Outside White*[1] and Waldby's book.

The Visible Human Project (VHP) is a team-effort enterprise to volumetrically scan the internal topography of the (dead) human body. By contrast, Horne's artifacts are a testimony for the body that *may* remain visually and tangibly present, after one's demise. Independently of each other, yet with subtle intertwinings, Horne's visual works and Waldby's critical text serve as an example of de-carnalizing the body, that is, observing it as a collection of objects. Both Horne and Waldby do not perceive the body as a commodity which we try to improve and perfect throughout our lives; somehow we have moved beyond this. Instead, we aim to reach the extreme objectification of our bodies, thus stretching beyond their carnality and achieving the impossible—immortality. "LbL" or "Life beyond limits"—is an acronym created to express the manifestation of a body that seeks to transcend its insufficiently designed mortal flesh.

Prior to the invention of artificial light, there was a clear, regulated distinction between night and day. When electricity became part of our lives, our circadian sense of space and time has been irreversibly affected. According to John O'Neill, "We hate to switch off our engines; lest we switch off ourselves, we have motors running, the lights on, the radio in the background. The more they [the machines] kill us, the more we turn to them for safety. Here's the very core of modern iatrogenesis" (179). Ironically, these machines reenact the Nietzschian killing of God (although this time they "kill" us), perpetuating a primeval need to destroy, shock, and then (if possible) re-create us and the world anew.

Although we depend on these machines, over the years we have determined that they have a better memory than ours, can store much more information, can engage simultaneously with many other machines creating networks, and can last longer than us. We have noticed that their inside, once broken, could be replaced to function unaffected by that change, if not better. These machines are updated on a regular basis, discarding their old components in favor of new ones.

If we have colonized the night, and, in a manner of speaking, outer space, we have not yet been able to completely manipulate our own territories, extending ourselves in space and time. Perhaps we should stop and "reboot" ourselves, only this time borrowing from the machines their characteristics. Irrefutably, we are not the machines, but no one can deny that between us there is an intimate relationship of love and hate, admiration and ennui. Because of this relation, or the "crossbreeding of bodies and machines" (Grosz, *Space* 110), now we refer to our bodies as being posthuman, namely, "a technology, a screen, a projected image" (Halberstam and Livingston 3).

Furthermore, it helps to note that in the 1990s, the National Library of Medicine and the National Institutes of Health started the Visual Human project, "[w]hose primary objective [was] to create the computerized equivalent of the anatomical atlas" (Thacher 169). Dead people's body were volumetrically scanned and then frozen. Waldby's book on the VHP analyzes in depth this controversial project which "[e]nacts the proposition that the interface between the virtual and actual space, the screen itself, is permeable, rather than a hygienic and absolute division" (5). She explains how the performance of dissections during the Renaissance generated the birth of the body-qua-text. She makes the distinction, however, that in the VHP no dissection whatsoever is performed, because the purpose of this project is to "transluminate the body" (25).[2] In order to do that, excriptive techniques are employed, techniques "[f]or imagining the living body [as] a visual 'dissection' which does not involve an actual incision" (90).

This section started from the premise that our bodies' structure is akin to that of a book, namely that both are readable terrains. Ever since we progressed from an oral to a written culture, we have been leaving proofs of our existence by using hieroglyphs, other carvings in stone, cave paintings, stylus on waxed tablets, ink on paper, and, more recently, keyboards. When we transluminate a body, we engage in a novel dialectics of writing and reading it, which determinates it to relinquish its intimate surface and, instead, constitute itself as an object of study. While our body is still carnal, sensual and palp-able, it opens itself toward a phase where space and time inhabit a different dimension themselves. The excriptive body via the visible human project tries to replicate the space-time continuum, thus not coercing itself anymore to fit the limitedness of its entropic laws.

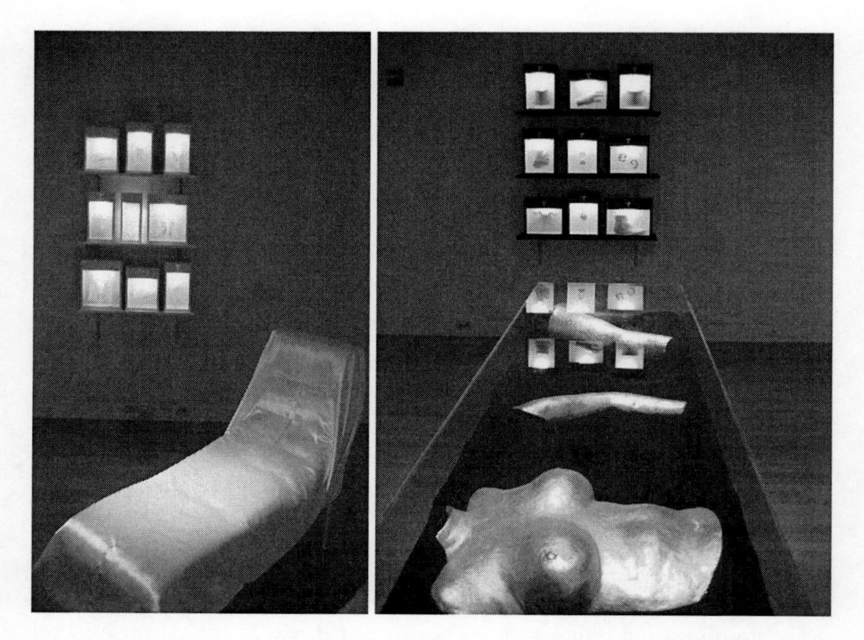

Figures 2 and 3. Natalie Horne. *Vital Necessities* and *Physical Accessories* (2005).

Consequently, this project denies the importance of the unconscious. Here "LbL" breathes intimacy with "BwO"—"Body without Organs." Bennett explains this concept as

> a multispecied and ongoing project of becoming in which new links are forged. BwO is the weird science of self-rehybridization. According to Deleuze and Guattari, BwO is a creature that hovers between human and nonhuman, between who-ness and it-ness. It is an assemblage. [...] BwO mimics the quivering dynamism of electron flows. (26-8)

In this context, "LbL" means having a body ready to be relived, revivified. Its who-ness or identity is derived from its necessity to continuously reinvent its it-ness. Following this scenario, then neither a body's who-ness nor its it-ness is ideal. Our next step in de-carnalizing our bodies becomes synonymously with optimizing them. Waldby asserts that we want to screen our inside and program our bodies through the "IatroGenic desire." Taken out of the medical vocabulary, yet nonetheless intended to be applied in medicine, too, the IatroGenic desire "[i]s the desire to create, not disease, but rather kinds of bodies which are stable. The iatrogenic desire could be summarized as the desire for programmable matter, for a capacity to order materiality according to the algorithmic efficiencies of the computer" (114).

If the VHP is still in its incipient stages, we have already started to "program" our matter via prosthetic devices. Initially known as "prosthetic limbs" which were made integral parts attached to the bodies of those persons whose arms and/or legs had been amputated, now prosthetic devices have extended their market as well as their definition (a pertinent example is the computer/artificial life).

Amputated, or even better, dismembered, the bodily parts in Horne's exhibit assume many readings. The human body is the new museum with pieces periodically remodeled or kept under a strict (cosmetic) diet. Or, as Toby Siebers points out, "The most noticeable change of recent decades is that [...] the body be exposed as the true and only subject of art" (5). He also remarks that the new art is "[m]ade of bodies [...] because that is what we care about. [...] [The new art] presents us with the death of the living, and the rebirth of something [...] beyond our control" (224-25). The represented body in Horne's portable museum exists amidst one central item: the clean, white and emptied bed, as if life were something that now we try to remember—life as a slippery memory (card)—and thus abstracted to its extreme. Or the bed could be read not as an intimate site anymore, but a dissecting table onto which, because we have not yet been able to program our matter, we have decided to sever our limbs, to destroy them. Dismembered, Horne's represented silver limbs, as well as other parts of her body, hang in the air, searching for a lover/curator/writer to reassemble them.

If so far we have conceived of the body *redivivus* as having the connotations revivified, relived, when adding Horne's vision into equation, we may define it as copy. According to Plato, a copy is twice removed from its original, hence less valuable. Let us keep in mind that a body is a symbol or a collection of symbols with mutable, (self-)eroding signs; however, its whole system is built on copying, transmitting information from one cell to another. Moreover, in a digital environment, we want to permanently multiply our physicality so that we could validate it *post mortem*.

If we go back in time, we notice that throughout the history of art, Claude Monet's Rouen Cathedral series (dated somewhere in the 1880s) will remain an innovation, with the cathedral changing its physiognomy in alluringly different types of light. What Horne displays is a collection of bodily copies. We deposit our sensations into an emotional bank whose codes are viscerally known by our carnality. However, there are instances when we need linguistic signs to mediate this transfer. Horne translates her body into bodily replicas. There is neither a verbal nor a carnal mediation there. We could best define her effort by saying that those bodily replicas stage a body that was at a certain moment in time. These bodily parts become her cathedral with suavely played *organ*(s) music. Conceived from this angle, Horne allows a body the possibility to contemplate itself and to literally touch parts otherwise physically impossible to be reached.

This type of intellectual intimacy reminds me of a very poetical note: Oscar Wenceslas de Lubicz Milosz's *L'amoureuse initiation*: "All these constellations are yours, they exist *in* you; *outside of your love they have no reality*! (Bachelard 189, emphasis added). A body *redivivus* is an overly poetical site, too, without end, justification, logic or fear. Outside its hinterlands, unplugged from its primordial senses, there is truly nothing palpable or desirable.

Point in fact, Horne's dismembered cast-in-silver members are a stylized lesson about the human body. Again, they may touch upon the Deleuzian notion of "BwO" as developed under the influence of Antonin Artaud's "Theater of Cruelty." The latter argued that, in order for the audience to respond better to a theatrical representation, each performance should have a necessary element of cruelty without which we would not be able to move beyond its entertainment pleasures. As explained in his *The Theater and Its Double* (1958), an element of cruelty, a novelty note masterfully placed amidst theatrical clichés, has the capacity to incite and/or shock our thinking and thus shake our prejudices about literature and, by extension, life. On the other hand, Deleuze envisions a body beyond categorizations whose boundaries and organs are open and mutable.

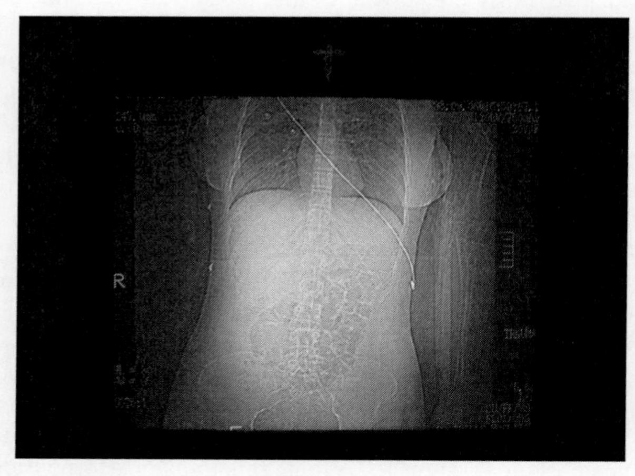

Figure 4. Natalie Horne, *Torso I (Detail)* (2005).

Figure 5. Natalie Horne. *Physical Accessories* (2005).

Extending this notion to our organs/bodily parts, there should be a method that could allow us to flow, go beyond our carnal screen, and consequently reinvent our bodies and their needs. Polished and shinny, Horne's dismembered bodily parts have covered over their imperfections, yet their bright color blinds us. Precisely because of their curvature, we cannot possibly see ourselves perfectly reflected, and, for a second, may even appear incongruous. Moreover, when one contemplates her work, one may think of silver as found in coins, medals, jewelry—all objects—and thus the body projected as object onto which the silver metal has melted. Could we read Horne's work as an example for *mise en corps* or *corps-e*? Is it about the alive or the dead? If it is about the alive trying to double its organs and limbs, could we stretch the discussion toward cloning? As Waldby remarks, "Medicine's task is to accumulate life-capital against the claims of death, so that the extension of life is understood to marginalize death to push back its limits through strategies of prevention" (141).

No matter how we decide to continue this discussion critically, we all want to learn quickly how to live on. Horne titles her work *Immortalis: Inside Green, Outside White*. The "green" that accompanies the "inside" could make one think about patina and statues, and how we try to preserve something that is immobile (which is a contradiction in terms, or maybe it is a transgression, a mutation of notions per se. If we would like to reinvent ourselves, our concepts should be reinvented, too). The "white" that modifies the "outside" brings to mind the bed presented earlier into discussion. Now her exhibit looks like an adult variation upon the Lacanian *le corps morcelé* (French for "the body in pieces"),[3] with the bed as the central item that is incapable of still dividing its memories imprinted between its luscious sheets. Then, if the inside is kept green, its outside may look white: untouched, unmodified and preserved—in a manner of speaking rendered immobile.

In addition, both Horne's work and the images proposed via VHP are a variation upon the old motif of *memento mori*. Speaking about the pathos of the VHP figures, Waldby notes that "[t]he image preserves the dead in the particularity of life and serves to remind the spectator of Death" (140). Ironically, this is something we have tried to escape from since we would like to share in or steal the palindropic temporality where, as Vivian Sobchack remarks, the object's surfaces could "[b]e read backward and forward without a change in meaning" (qtd. in Waldby 130). Hypothetically speaking, we achieve this so long as we are not consciously aware of our capital temporal frames of "before" and especially "after" our lives. Horne's exhibit achieves this, too, when we realize that the stylized, silver right limb could not be distinguished from the left, because they have not been marked, labeled as such. Only the interior of the represented body is excriptively illuminated and studded with jewelry. After all, life does matter, although its matter gradually disappears into thin air, which coincidentally is the final endpoint of the VHP: "If the human body could be stabilized as mechanical system, the medicine's interventions could potentially return the diseased living body to original conditions, a return to equilibrium, and the problem of death could be postponed" (Waldby 130).

Until we achieve this ultimate/supreme ex-tension of our bodies, we may continue to imaginatively design the aesthetics of the body *redivivus*, desiring to find the trick to being alive. The question that remains unanswered is: who will then be tolling the bells for us? To answer this, in Morris Berman's *The Reenchantment of the World* (1981), he reminds us that "From the 16th-century on, the mind has been progressively expunged from the phenomenal world. […] The logical end point of this world is a feeling of total reification: everything is an object; […] and I am ultimately an object, too" (16-17). From that 16th century perspective, we have moved to its extreme, wishing to exhibit our possessions, including our bodies. If we aspire to possess an extroverted dimension, then this reasoning could reinforce the ideas promoted by the cosmetic industry. Horne is interested in seeing the body from a physiological perspective, namely as a system of networks. To the "[s]phygnoscope (a device for rendering the pulse visible), the myograph (an apparatus for recording muscular contractions), and the cardiograph (an instrument for tracing the heartbeat) [that] were crucial […] to the scientific disciplining of bodies and communities" (Cartwright 12), Horne adds her perspective on the body consisting of portable sculptures and excriptive-like images, thus enriching an otherwise general view of the human body.

Figure 6. Natalie Horne. *Vital Necessities (Detail, Heart)* (2005).

Since our bodies are the fluid metaphors par excellence, discovering their final referent/embodiment is similar to our perception of the horizon, whose lips are constantly misleading and receding. The body experiences an effervescent folding, pouring itself into the future while keeping/storing the echoed versions of its former embodiments.

Nonetheless, if the body and mind experience several variations of their manifestations, is there anything from our sentient embodiments that somehow possibly continues after death? We definitely live our bodies, but do they have at least a story that is continued after their demise? Like Marcel Proust's famous *madelaine*, let us remember those stories read in our teen years; they had an end, but once we started rereading them, we repeated the story's narrative over and over again.

The questioned repeatability of our bodies and minds is expressed in the manner in which we perceive time. Because today we live under the protean sign of the electronic swift and stackable spatio-temporality—necessarily added to the former discussions of the nature of time (linear/cyclical; centripetal/centrifugal)—the latest discourses refer more and more to its entropic/negentropic duality. Sobchack believes that "The electronic instant […] constitutes a form of absolute presence. […] As subjective time becomes experienced as unprecedently extroverted […], space becomes as abstract, ungrounded, and flat—a site (or screen) for play and display" (158). But what does "absolute presence" mean? Is it a surrogate for our still searching absolute freedom and embodiment? In the electronic dimension, the screen/interface manipulates the image we have of our bodies, but not our bodies' constitution per se. The screen is flat and functions like a portable prosthetic mirror transporting us from one volatile representation to another. With a touch of the keyboard, the body's image could be wound backwards, instantaneously experiencing flashbacks, while still residing in the present. It seems that the VHP has pushed this idea even further, proposing that a dead body's topology could be scanned to better serve the medical endeavor in finding possible explanations for the etiology of some illnesses, and thus propose prophylactic amendments. Eugene Thatcher contends that for digital anatomy,

> The complexity of the proposed intersections between the patient-body, medical technologies (X-ray, CT, MRI), and new computer technologies moves toward approaching the biomedical body as a field of informational diagnosis, where what is to be trusted is not […] one's own eyes […] but rather an informational calculative dimension to the body and materiality. […] [d]igital anatomy is concerned with how to technically procure the most analytically optimized body technology configuration possible. (181-82)

Actually, this kind of body technology reveals a deeper layer within the VHP where the "[p]alindromic temporality is so attractive for the biomedical imagination because it plays out the IatroGenic desire for a body which acts as mechanical, rather than chaotic system" (Waldby 130). Science has its own elaborate visions for the optimized function of the human body. Put differently, dreams are not the royal path to one's unconscious anymore; nor, in Jungian terms, are they an individual's "compensatory activity rather than a disguised wish" (Eberwein 14); nor do they result exclusively from the mind/body isomorphism. Science itself has dreams *and* fantasies it projects onto us.

Is Horne's exhibit a complement or an attack to that branch of science that would like to keep us intact, barely touched by age and physical decay? Actually, the uniqueness of her work comes as a shock, when we discover that her artworks are accompanied by a cynical pamphlet.[4] If the artworks propose a poetical view of the illuminated bodily within, the pamphlet is pertinent and irrefutably actual. In a bold manner, it addresses the classical theme of *vanitas*: we still seem to care a great deal about our bodily façade. The dichotomy presented in Horne's title's work—"inside/outside" with its chromatic equivalents "green/white"—presents a clue. We work hard to keep our inside green so its outside remains shiny, white. But our bodies have their own dynamic and a recorded tendency to direct themselves "towards simplicity and quiescence, impelling [them] towards death. Life can be seen, on this Freudian scenario, as the limited deferment or delay of the death drive, a detour of death through the pleasure principle" (Grosz, 201). On Horne's pamphlet, there are two symbols: of infinity and a stylized version of the serpent coiled on a rod (typically used in medicine). Both imply a circular, loop-like, physically and mentally exhausting movement.

Describing the products fictitiously marketed by *Immortalis*, Horne writes: "Health—it's newest accessory! Show off your vitality with custom-made, limited edition replacement organs, prosthesis parts, and multi-treatment pharmaceuticals. Allow yourself to experience life forever—in style!" Who has access to this life in style than those with fortune? Consequently, the adjective green from her work's subtitle is decoded as referring to the American dollar. Furthermore, as described in "Our Guarantee…" section of the pamphlet, "The concept of *Immortalis* derives from a fascination with a material culture. […] The most common type of exchange is things in return for money. Even where health and human life are involved, the transaction of 'things' is still prevalent. […] Simply maintaining or being grateful for life we have is a challenge for many, yet for some it will never be enough." The concept *immortalis* could not possibly apply to everyone since there are restrictions when we think of what makes us immortal or who will attain this privileged status.[5]

Then, to return to Waldby's quote from the beginning of this section where she contends that "The anatomical body is co-emergent with the anatomical text," we realize that, since we cannot all secure our immortality, or drastically and periodically revision our bodily components, then maybe we start learning how to read the newly released books on/of our interiority. We are studded with receptors for pain, pleasure and other sensations; perhaps, we may insert a receptor for curiosity that could keep us updated at the changes that occur gradually within our revised interiority. Otherwise, we will maintain a damaging hiatus between the science of our bodies and their literature, denying our knowledge to expand, something that could be just as good as (if not better than) immortality. Unlike computers, we cannot really afford do develop a deficit/malfunctioning of intimate knowledge.

Félix Gonzales-Torres

One space spreads through all creatures equally—
inner-world-space. Birds quietly flying go
flying through us. Oh, I that want to grow,
the tree I see outside grows in me!
(Rilke 193)

In many of his poems, the German poet Rainer Maria Rilke employed the Weltinnerraum motif [German for "inner-world-space"]; the term refers to a world situated beneath the epidermis, the introverted world of the self that is known to us instinctively. For the Cuban-born artist Félix Gonzales-Torres, his Weltinnerraum becomes part of an art of loss or an aesthetic of being at loss. Hence, it becomes a representation of an unstable art, in constant desire to dream of the lost lover while anticipating his own death. The artist died young, when he was in his late thirties, from complications resulting from AIDS (his partner, Ross, had died earlier of the same illness). His artistic motifs revolve either around the theme of double eulogizing his partner's death (e.g., these works contain two items: clocks, curtains or pillows), or around a multiplying, never-ending supply of items (e.g., candies or papers). His oeuvre questions the meaning of artworks, as well as the autocracy and necessity of still exhibiting them exclusively in museums (In this respect, I discuss one of his billboards shown outside of a museum jurisdiction).

Art historian Siebers believes that art has been fascinated with the human body and that "[t]he human history is wedded to the mystery of the body" (3). But when the body is not represented, as is the case in Gonzales-Torres's works, could we agree with Siebers when he proclaims that "There is no perception outside of the body" (1)? When the mollusk that created the shell is gone, does this mean the shell does not exist anymore? Gonzales-Torres's aesthetic defies this idea that limits the topology of one's embodiment, proposing the continuity of the human body in objects, memories and people.

To start feeling this artist's works, it is useful to remark that they are "constantly dematerializing sculptures" (Storr 8), and the best way to exemplify this is through one of the artist's most famous and controversial works, *Untitled [Ross]* (1991). Since we are referring to these works as being in constant (and literal) re-creation, then it is important to note from the beginning that there have been uncountable variations to his installations. Some variations of this particular work have been titled *Placebo*, an important motif for this chapter. I have chosen the one dedicated to Ross for several reasons. First, it already contains the placebo motif; second, it is dedicated to his lover; third, the stack of candies is placed in a corner (in other installations, the candies were left on the floor in a rectangular shape); and finally, the installation contrasts effectively the series of dichotomies of inside/outside as well as of museum/nature.

In addition, the candies are arranged in a temporary, fragile pyramid that conveys a feeling of immobility. For Bennett, the term enchantment "[e]ntails a state of wonder, and one of the distinctions of this state is the temporary suspension of chronological time and bodily movement. To be enchanted, then, is to participate in a momentarily immobilizing encounter; it is to be transfixed, spellbound" (5). At first reading, how could one participate in something that does not require action, but is defined as a momentarily immobilizing encounter? Isn't action a sine qua non condition of participation? Through this series of questions I reinforce an idea that pervades Gonzales-Torres's aesthetic; he is not afraid to become immobilized now and then, or to imagine the body (his own or his lover's) as momentarily taking non-carnal shapes/forms. By so doing, he becomes part of a reality that is boundless, infinite and hence regenerative.

Interestingly, in this photo of the installation it seems that the tree—and not the candies—is inside a closed glass, nonetheless still receiving its energy from within. Another admirer of nature, the Japanese music composer Tōru Takemitsu, writes in his memoir, *Confronting Silence: Selected Writings* (1995):

> Immobile, they [the trees] show that repose with arms, palms, fingers—like Buddha. [...] In a magnificent way trees transform time into space. Geometrically precise, their inner growth rings gradually expand in time to fill unlimited space. [...] In undivided action and with a glance toward

> infinity and eternity, leaves create chlorophyll, roots absorb minerals. (130)

We, too, refer to some of our growths as rings, only for us they are not metaphysical, but unfortunately physical (for example, we notice how our bodies develop fat rings, wrinkles, etc.). In so doing, we keep ourselves rooted in a physicality that is more and more corrupted by the passing of time.

Now it may be clearer why this work in its other installations has been titled *Placebo*. I admit that, without knowing anything about Gonzales-Torres's intimate concerns and artistic identity crisis, I would inevitably fall into the mistake of reading these candies as mock-art pieces. How could one viewer take a candy, eat it, and consider s/he was admiring, or was in the presence of an artwork? Actually, was that viewer in the presence *or* in the absence of an artwork since—by eating it—the work instantly disappears? Without the viewer knowing, Gonzales-Torres's effective aesthetic comes into action. By ingesting the candy, he makes us participate in his art, or continue his artistic gesture. The contact of the candy with our mouth represents a variation on the theme of enchantment, a literal encounter with the other. The candy tickles our taste buds, is dissolved by saliva and becomes assimilated by our organism.

Nonetheless a candy is a sugary simulacrum for another need, and in this context, eating it does not reciprocate the lesson of unconditional love and sacrifice offered by the Eucharist. We eat our candy alone, individually. We become thirsty and/or addicted to continued eating. For Simon Watney, when this work is read in the placebo-key, it immediately

> [i]nvolves us in the cultural field of the medical clinical trails of potential treatment drugs. A placebo is an inert substance, indistinguishable from a pharmaceutical compound in comparison to which the effects of the drug may be measured. [...] And yet a placebo is never just an inert substance, for it inevitably carries with it a profound supplement of hope. (159)

Just as before when we encountered the paradox of participation in absentia, we are challenged by another confrontation due to the puzzling idea of "inert substance." To Watney's explication, I add that, in the light of Gonzales-Torres's artistic philosophy, our bodies, relationships and jammed past-present-future contain a latent inert substance through which we are reminded that we can achieve only partly the full potential of our bodies. Furthermore, there is an ambiguity in the word "Placebo"; ad litteram, it is the future form of the Latin verb placeo, i.e., "to please," but its referent is always equivocal. Bennett argues that "The human body possesses an astonishing network of relays between, for example, human thoughts and arteries. Evidence of this network is provided by the placebo effect" (170). However, one feels entitled to say: "Placebo" = "I will like...," but what will I like? Undoubtedly, the verb has deep or loose connections with desire, and yet, because the desired object is missing or ambiguous, the construction continues the series of paradoxes mentioned thus far in the chapter. One should find it useful to remember that "[r]eality is paradoxical, that things and their opposites are closely related, that attachment and resistance have the same root. We know this already on an intuitive level, for we speak of love-hate relationships, recognize that what frightens us is most likely what liberates us" (Berman 80).

In the context of Gonzales-Torres's work, "Placebo" could result in the following translation: "I will like the candy," "the show," "the artist's vision," "my liberty to refuse to eat the candy," "the desire to throw it," etc. Soon we realize that we have just entered the artist's game which is directed toward mutability and never-ending musings at ourselves as the otherwise latent other. Furthermore, Storr reminds those who never saw Gonzales-Torres's installation with the candies that the artist specifically exhorted the curators that "no effort should be made to replenish the stock of candies [...]. To restore what has been taken or formally correct the aleatory effects of multiple acts of removal would be to interrupt and administer a spontaneous social transaction" (7). He did not want the stock of candies to be re-supplied because—like a synecdoche—they stood as part for (the whole of) Ross. By not replenishing the candies, Gonzales-Torres reminds us that people's lost beloveds are irreplaceable. From a different perspective, because there will be another installation with a new stock of candies, the artist is also aware of our contract with loss. We have intervened even there to suggest how much time (usually little!) we should mourn a lost beloved. Our society does not allow us or recommend that we have a prolonged period of mourning, for a mourner may thus be labeled as pathologically melancholic and not suited to control his/her emotions and confront reality. However, the act of mourning is an intimate moment whose intensity and duration should not be directed from outside. Therefore, society's discourses cannot ever function as the panacea for an absent lover.

The theme of mourning is also suggested in *Untitled [Perfect Lovers]* (1987-1990), where two ordinary, round clocks are placed next to one another. Both clocks point to the same hour; one hand "reads" the time when Ross died, while the other when Gonzales-Torres's identity as a lover, partner and human being was shattered in pieces. Lewis Bautz describes this work as questioning the "idea of portraiture, 'immortalizing the sitter'" (214). Interpreted in this key, we could also think of tombstones—immobile and transfixed in a reality that does not change, thus opening the path for the eternal duration. In addition, tombstones and Gonzales-Torres's clocks give the impression of a time whose real ticking is continued through the internal site of a lover's accumulated memories. Through this vision, the lost/absent other becomes idealized, just like memory, for Henri Bergson, "[i]s nothing other than an *expansum* of dimension, the expansivity of which is internal, immanent, and intensive—and, consequently, will not be truly spatial, but rather ideal" (Martin 66).

Gonzales-Torres's work permits yet another interpretation by raising a series of questions: When are our bodies completed? When do their internal clocks stop ticking? Could the lost/absent lover's body be reanimated through memory, and thus—like a mechanical clock—periodically be wounded over again? Remembering the dead probably achieves its ultimate form through acts of repeating the process of abstracting the one who passed away, but who nonetheless continues to live in the lover's memory.

If that is a possible reading of a longed for, absent lover, we could reinforce this idea through a complementary work, *Untitled* (1991). The bed is empty and it feels like the ones who slept in it just awoke and did not have time to tidy it up. Or, knowing the symbol of double-ness that is quintessential to Gonzales-Torres, the whiteness of the nocturnal ensemble (consisting of two pillows and one bed sheet symbolizing or etching in intimacy the couple's united bodily contours) is a more personal equivalent of the already discussed pair of clocks. The intimacy of this couple reminds me of a process called "Pelastration" that occurs in the theory of "The Big Tube." In that case, "[a]n impact force penetrates an infinite stretchable flexible layer of a space and carries a part of that flexible layer as an additional outer skin during all [its] further trajectory" (Laureyssens). Death has such an impact in our lives projecting us into an abyss from where, cathartically, the demised one becomes part of our now extended self. After he has lost Ross, the artist realizes that he has become his lover's skin, containing and/or continuing him in memory. Nonetheless, bereft by pain, Gonzales-Torres is tossing and turning in an empty bed, embracing a hollow space, and thus the memory of the absent lover eventually exhausts him. Death has no soothing explanation, and we try "[t]o recover from *the theft of meaning*, [but] everywhere we look we find only the heartless fact of finitude. [...] Acceptance of the finality of death induces humility—humility in the face of the Other and humility with regard to the possibility of knowing anything with certainty" (Bennett 78-79, emphasis added). Not only do we live (through) a body-without-organs, but we also face dealing with the unpredictable cycle of memories that hurts, heals and re-opens the wound of having lost our beloved.

Then, probably, science is not the only site where theories become antiquated, and throughout our lives we frequently change our self-perceptions. The fundamental difference is that, unlike in science, we do not seem to abandon these antiquated theories/views, but instead let them become the residual material of our dreams, fantasies and nightmares. We also do not abandon the memory of those who died before we were born. For example, Plato, Confucius, Shakespeare, Borges and Foucault died (just to name a few famous thinkers), but we still use their ideas and by so doing we reanimate their existence. Although they are physically absent, we make them real in our intellect and imagination and confer upon them a posthumous existence. Consequently, a voice may not seem useless only because it is disembodied. If this is so, then how could we conceive that our beloved ever really dies?

Returning to Gonzales-Torres's works, the bed is white and deserted; the clocks do not indicate a palpable, real time; and the candies are thrown on the floor. Is there anyone present in his works? Or are they the supreme representation of the other's absence? The artist invites us to sift through the pain of the loss, to continue to dream about the absent other, and to make him/her exist transcendentally. In a manner of speaking, we realize that our fear is not so much about the other who died, as about our own reluctance to continue him/her in objects that s/he touched, adored, was fascinated or intrigued by. Gonzales-Torres's project seems Pygmalion-like, although with a more refined, never fixed-in-one-form Galatea. He tries to encapsulate the (last drop of) essence of his absent lover, which could also stand as a metaphor or *metaembodiment* for the next piece, *Untitled [Petit Palais]* (1992). This work typically consists of "24 or 42, 15 or 25 watt white light bulbs distributed evenly over several meters of electrical cord" (Bautz 215).

Gonzales-Torres plays with another concept here, that of light/illuminating source, and at first glance it seems a little eerie to look only at the bulbs, instead of what light usually offers—the presence of the other (objects/people). To better understand this particular work, let us analyze it in conjunction with air, which "In the world of modernity, has become the most taken-for-granted phenomena. Although we imbibe it continually, we commonly fail to notice that there is anything there. We refer to the unseen depth between things—between people, or trees, or clouds—as mere empty space," yet the air is "[t]hat element that we are most intimately in" (Abram 258-59). As with the air, when it comes to light we are trained to see something.

However, because all that we see in Gonzales-Torres's work are the light bulbs, we may pass indifferently by them. Unlike Zen philosophy, most Westerners are not attuned to seeing the benefits of an empty space, or—better yet—when we see empty spaces we think erroneously that that is all that we should see. This particular work, as well as its several installations, reminds me of the American Realist painter Edward Hopper's use of light and darkness, of the angular spaces where light projected itself, of diners with dim lights and consumed sitters, and of those paintings where one viewer wondered what was inside a house with barely lit rooms.

Proposing an opposite interpretation to light, Gonzales-Torres exposes only its bulbs, in an attempt to make us finally see the illuminating source; the work shouts "Let there be uninterrupted light!" He does not use light to play with dichotomies—on/off, dim/bright, natural/artificial or intermittent/constant. His work is enveloping, catching us staring at its nude presence. Placed next to the art of loss, these lights could be interpreted as candles that keep vigil through the night and remain as company to someone who is mourning. Therefore, the light becomes a warm presence. Just as the natural light possesses a wave-particle duality, so this particular work brings the comfort and discomfort of confronting difficult moments. Because the bulbs of light are not accompanied by any other object whose contours could be gently revealed and caressed, when we see this work we are face to face with our own thoughts and become introspective. As Lewis Bautz suggests, "[o]utside of a museum or a gallery context [the lights] resist identification as art, the light strings are [the artist's] ultimate gesture of implicating the perceiver in the construction of meaning" (215). I am not sure if the lights per se constitute the artist's ultimate gesture in reaching the final meaning of his artworks, taking into account how art has traditionally been constructed and displayed, keeping the viewer at a distance. I think his whole body of works challenges our (mis)understanding of artworks. For example, the billboard discussed earlier has been exhibited outside, made available for the lay community, been caressed by the rays of light and touched by the drops of rain. The artist wanted to devalue/diminish a museum's too closed/selected audience, but, since a museum is one establishment for power, it will always have an authoritative, controlling voice.

Another intriguing aspect about Gonzales-Torres' works is their untitled status. Part of the explanation for why they were left "untitled" was apologetic on the artist's behalf. As a Cuban-born artist, he thought that naming artworks in his adoptive language would be inappropriate. Furthermore, labeling them would mean ascribing to them a limited word combination. His series of untitled works remind me of what the famous Italian cineaste Federico Fellini once stated: "To give unambiguous final answers is not impossible—rather, it is immoral" (qtd. in Kovács 165). Gonzales-Torres's untitled works also make me think of René Magritte's painting *The Treason of Images* (1928-29), where the Belgian artist paints a pipe accompanied by a very explicit phrase that reads Ceci ne pas un pipe, i.e., "This is not a pipe" (meaning it is more *or* less than a real pipe, and thus more than we usually see and/or words typically say). All these artists affirm that we rarely make an effort to discover latent truths/versions underneath (complex) situations. In this context, Gonzales-Torres's *Untitled* (1990) stands as a good example. It consists of a stack of red papers (other installations consisted of stacks of white or printed paper). As with the work with candies, visitors/viewers are encouraged to take one paper as they leave the museum. About this artistic gesture/offer, Russell Ferguson comments that "The print sheet of paper is worth nothing in material terms, but in other intangible ways it is invaluable. [...] [w]hat price can really be put on this gift that has to connect one intangible sensibility with another? On the other hand, since the stacks are theoretically endless, what monetary value could possibly be placed on one element of a potentially infinite series?" (84)

Gonzales-Torres's multiplied/repeatable elements have been interpreted alongside Walter Benjamin's idea according to which—exposed to mechanical means of reproduction— artworks have lost their aura in modern times. Years have passed since Benjamin's theories, which were stimulated by photography, were first presented to the public, and there have been recorded many technological achievements. Yet, to accept that we benefit from Xerox and digital machines, cellular phones and other technologically sophisticated items is rather counter-productive since they create in us a feeling of uninterrupted being-ness.

On the other hand, Gonzales-Torres's works are ineffable in implying that our bodies have a stretchable dimension underneath their limited material, yet decaying form. Within this ineffable and immaterial chrysalis of our material bodies, time knows and embraces simultaneously its extended and extending dimensions, its centripetal and centrifugal elements and its already formed (or should I say past?) and forming seconds. Thus, we discover that within the ineffable, there resides the enchanted moment of being. The stacks of paper represent a metaphor for the multilayered strata of our embodiments that unfortunately have become the "yesterday" and the "day before yesterday" of our body. The artist often remarked that we experience an explosion of information at the cost of an implosion of meaning. Alongside this socio-cultural trend, we do not know how to immortalize our lives: "Because people just do not remember, especially in our culture where we have an explosion of information, but an implosion of meaning. It is like *Casablanca*, when Humphrey Bogart said, 'A long time ago, last night.' People don't remember last night" (qtd. in Storr 30). To criticize the flaws of this culture, yet still to use its materials as examples, represents the brilliance of this artist's oeuvre. No one can criticize anything accurately unless one has seen it from within, and lived enough through that experience to understand it, and then mock it.

Gonzales-Torres's art consists of perishable materials and concepts that seem to be aware of their limited temporality. Yet this art is not about forgetfulness. Instead, it proposes to make the viewer feel lost, transfixed, contemplative, in love, denial or despair for the fleeting second. Because he builds his artworks on these perishable materials, Gonzales-Torres envisions a body that could be as versatile as these materials, thus belonging to many entities. Ann Weinstone contends that

> To belong refers, simultaneously, to belonging, belongings, and belonging to. This move breaks the frame 'individual,' a frame that remains axiomatic. […] To belong introduces an uncertainty coupled to a movement in time, a movement of deferral. […] The general operational mode of belonging is suspension, a holding of things together in such a way that they may move in and out of states. […] To belong is suspended with respect to its possessions: what I yoke is what passes through me is what I shelter. (28-29)

It seems that whatever passes through Gonzales-Torres's aesthetic is a vision of a body belonging more to the other (Ross, the artist's admirers and critics) than to himself.

Undoubtedly, within his artistic gesture there is kept captive the grain of despair as a result of these deeply annihilating-driven desires. About the installation with the candies he writes: "This work originated from my fear of losing everything. […] This work cannot be destroyed the way other things in my life have disappeared and have left me. I destroyed it myself instead. I had control over it and this is what has empowered me" (qtd. in Kwon 301-303).[6] To understand the idea of the artist negating his own creation, one must remember that its idiosyncrasy resides in its being a process, and hence destroyable, or potentially being recreated ad infinitum. These works are installations that are defined by this mutable, teasing signature. Deleuze's and Guattari's project of "BwO" becomes the perpetual-installing body. Nothing succeeds in staying/remaining in situ in an installing embodiment, which is probably a good definition for the antonymic pair Eros-and-Thanatos. Read in this key, Gonzales-Torres's works are simultaneously about love and loss, love and despair, love and destruction.

However, looked at more thoroughly, they intrigue the viewer by asking him/her to reflect upon the following questions: Are our desires linear, or intricate/convoluted and thus too complicated to be visually expressed? Who could assure us that our desires do not look exactly like Gonzales-Torres's repeatable, multiple objects that have been molded from on original matrix? Perplexing things intrigue and cha(lle)nge us, yet when it comes to our bodies, for one reason or another, we still prefer to perceive them like statues mounted on a pedestal: clean, undisturbed and supposedly completely "natural" (organic). I think the time has come for us to realize that there is no such thing as "natural" when it comes to bodies. Bodies, or even more accurately, bodily images are installations constantly constructed and deconstructed for what appears to be an aesthetics for an art for the body's art's sake (meaning those changes that are performed on our bodies to make sure they fit into a certain image we have created for ourselves, disregarding or disrupting the alterations that inevitably affect our bodies, whether they result from a chronic pain or as signs of aging).

Then, when it comes to bodies and how we perceive them, do we want to operate under an *anesthetic aesthetics* of trompe-l'oeil (and thus be deceived into believing their "naturalness")? Or do we want to accept their prostheticity, that is to say their constantly assembling, disassembling, entangling, and disentangling their bodily and mental components (up to the point they may become as confusing and difficult to label as Gonzales-Torres's artworks)? Our desires cannot possibly be fully understood, (re)stored and ultimately materialized. Gonzales-Torres has refashioned our notion of the ubiquitous perception of the body that is so much in view, yet, paradoxically, barely taken into consideration. We focus on our posture, habits, accessories but not so much on our needs. He suggests that our bodies are mercurial in nature, and even when they are no longer physically present, nonetheless are sensed through the evidence left on objects and people. His art also questions the functionality of our past: Who could move beyond the past? Where is one's point terminus? Because there is no past in the past, we may infer that we do not have an actual, traceable end either. If the legendary bird Phoenix is reborn from its own ashes, Gonzales-Torres enacts this motif by "recreating" the lost lover from the shared, consummated-in-intimacy, now ash-like past moments (offering one model for other people to "recreate" his protean identity once he is dead, and thus showing an alternative to coping with losing a beloved). Finally, the verbal image employed at the beginning of this section, Rilke's tree growing (and glowing) from within, could be decoded as those intimate moments that multiply inside us to branch out indefinitely in others later on.

Almodóvar's *All about My Mother*

At a certain point in the body's
entropic movement,
subjectivity drains away, beyond a
point where it
can be recaptured by technical
intervention.
(Waldby 162)

In an interview with Peter Canning, Deleuze asserts that "The concept of the image in the present only applies to mediocre or commercial images. […] The present is not at all a natural given of the image" (372). This is explained by the fact that cinema is characterized by a montage of flashbacks, creating the illusion of an extended present in which no one makes an effort to reel back anything since the past is already closely attached to the present. Deleuze has also argued that the human brain requires an interval in order to meditate upon an event's significance. Therefore, because our brains need a short delay to grasp the significance of events, the brains do not belong exclusively to the present either. In *All about My Mother* (1999), Almodóvar proposes the opposite by anchoring the story and its characters in what seems to be an engulfing present. For those who cannot hold their disbelief in suspension, or seek in a film only its superficial, entertaining pleasures, this film is impossible to watch. I had some difficulties myself because of its chaotically entangled scenario. To understand this type of scenario which does not seem to have any coherence and/or explanation, it helps to offer a scientific background. As Bennett argues, "In chaos theory, a particle's motion is said to have a strange attractor if the path of its transits is always unpredictable" (17). Almodóvar's film could be defined as an amalgam of kinetic brilliance (everything moves really fast, practically without stop) and an unpredictable, vertigo-like series of events.

The film starts with a close-up of glucose bags mounted on IVs in a hospital's Intensive Unit Center. The patient, whose identity remains anonymous, dies. Manuela, a nurse and one of the major characters of this film, leaves the intensive care room and rapidly enters another. There, she picks up the phone and dials the number of an Organ Donor Organization, telling the receptionist that the hospital may be able to offer a transplant.

Immediately after this opening frame, we see Manuela again, this time at home with her son, Estéban, ready to watch a movie starring Bette Davis. The clip shows an actress bored and annoyed with her fans, whom she believes to be vandals, uneducated and trivial. The clip subtly bridges the narrative in Almodóvar's film. Estéban is a young man, passionate about the arts— film, theater and writing. For his upcoming birthday he requests from his mother three gifts.

The first is to observe her in a simulated seminar session dealing with a bereaved woman who tries to learn how to cope with the loss of her beloved. This session is part of her training to become a better liaison between those who lose a beloved and those who may benefit from organ transplant. The scene is brief, precipitate and without closure, like everything else that has happened so far in the film. Almodóvar accumulates detail after detail in his film's narrative until it reaches a stasis, after which it will continue an uninterrupted avalanche into the theme of present/presence.

As a second gift, Estéban goes with his mother to see his favorite actress, Huma Rojo, performing the role of Blanche in Tennessee William's play *A Streetcar Named Desire*. After the performance, mother and son wait in the rain for Huma to request an autograph from her. While waiting, Estéban asks for his third gift: he wants to know everything about his father. Because Huma gets into a taxi so rapidly, she does not sign the autograph for Estéban. He starts running after the taxi and gets hit by a car. This is the moment that becomes the stasis of the film and constitutes a kind of minimal reference frame between the temporal segments of "before" and "after." The next frame depicts Manuela waiting for the doctors to confirm her worst fear, that Estéban is dead. She also finds out that her son's heart is not yet dead; would she be willing to let it live or die? In an entry written on the eve of his death, as a marginal note to a photo Manuela showed to Estéban from her youth, the latter wrote in his diary that he wanted to know everything about the missing half of the photo. This desire incites Manuela to start searching for Estéban's father and thus reattach the missing half to a heart that continues to live in different persons (in his mother because of the filial liaison and in the anonymous patient because of the heart transplant). Estéban's last wish is about discovering the mysteries lurking in the controversial concept of absent presence that keeps haunting us. According to the director, the concept of absence has been misinterpreted: whatever/whoever is not present is not necessarily completely absent as long as we cannot stop thinking about it/him/her.

After this long and intricate enumeration of ideas and events, we may be able to better understand why Almodóvar wrote such an uncanny scenario: Manuela loses her son in a car accident; trying to honor her last promise to him, after Estéban's death she starts searching for his father. In her search, she reconnects with Agrado, an old-time friend; manages to find a temporary job as an assistant of Huma; and meets Sister Rosa, who has AIDS and is pregnant. (Manuela discovers that Lola is also the father of Rosa 's son.) According to the director, "This atypical family [and scenario] evokes […] the variety of families that are possible in these times. If there is something that characterizes the end of the 20th century it is the rupture of the traditional family: now you can form families with other members, other ties, other biological relations that need to be respected" (qtd. in D'Lugo 102-103). In this scenario, there is also Agrado's unforgettable episode where s/he professes belief in the authenticity of a perfect*ly* dreamed body. Describing the parallel between films and dreams, Eberwein notes that the former "[a]ppear to us in a way that activates the regressive experience of watching dreams on our psychic dream screens. The actual screen in the theater functions as a psychic prosthesis of our dream screen" (192). It is important to emphasize that in this film no one is nostalgic about the past, and consequently no one has time or even desire to dream indefinitely, live a dreamy life, or refuse his/her authenticity and existence.

Furthermore, because in this film there are either biologically born women or transvestites (half impersonating women)— including Estéban's father, Lola—I propose a rereading of the legend of the Minotaur which may provide some insight into this film's too chaotically presented ideas. Luce Irigaray notes,

> The labyrinth, whose path was known to Ariadne, for example, would thus be that of the lips. This mystery of the female lips, the way which they open to give birth to the universe, and touch together to permit the female individual to have a sense of her identity, would be the forgotten secret of perceiving and generating the world. (qtd. in Lorraine 32)

It seems that in *All about My Mother*, the women are caught in the maze, too, suffocating themselves and the Minotaur, searching frantically for the thread that will liberate them. The director finds a satisfactory solution to this problem only after Manuela has honored her dead son's wish. This is the moment that seems to release the *last Minotaur* from the labyrinth (which stands as a symbol for our inhibitions, as explained later), and offer some kind of closure to the film.

Moreover, of all these complicated characters whose identity is at a borderline, Agrado plays a quintessential role. She describes herself, saying: "They call me La Agrado [Spanish for "Amiable/Approachable"] because all I want to do is make life agreeable for others. Besides agreeable, I am also very authentic." Almodóvar proposes a subtle analysis; on the one hand, there are practically no men in his film. Even those who are present speak only short lines that do not contribute significantly to the film's narrative. As Marvin D'Lugo notes, "In the intricate weave of surrogate and refigured identities within the family, patriarchy is resemanticized, principally through Lola, the man who would be a woman, the woman who is the father of both Estébans" (102). On the other hand, the idea of the organ transplant presented briefly during the opening frame of the film functions as a means through which the director analyzes the current concept of being authentic. When Agrado claims that s/he is "very authentic" and proceeds to detail it, we find out about her/his several cosmetic surgeries (As a prostitute, s/he needs to have an "authentic" body, without which s/he would be out of the meat market). She concludes her monologue by saying that "A woman is more authentic the more she resembles what she dreams herself to be." Almodóvar rewrites some existential/ontological theorems; we do not have a body, nor do we belong to a body, nor do we live a body, as much as we dream to a body which approaches authenticity the moment it gets closer to our internal fantasies.

Having said this, let us return to a moment in the film that is presented only briefly, the organ transplant. Because this film is about living, the director takes the liberty to accept/be tolerant toward all forms of living: Agrado has had several cosmetic surgeries (which could be interpreted as external or epidermal transplants); a long time ago, Lola decided to be a woman in a man's body; Sister Rosa has sinned and had sexual intercourse resulting in her current pregnancy; sick and desperate patients wait to receive organ transplants; and Manuela convinces herself to continue her life if only she honors her son's last wish. Almodóvar asks us not only to accept family ties created outside of the traditional pattern, but also to educate and diversify our rather limited views on tolerance. If we have difficulties accepting other people's personal lifestyles (and here is a parenthetical observation on his own homosexuality, too), a lesson given by anatomy always serves as a good example. More explicitly, in order to survive, a body that is under surgery and receives an "alien" organ must exhibit a considerate amount of tolerance. Renée C. Fox argues that

> Incorporated and assimilated into its scientific lexicon are notions of the 'tolerance,' 'acceptance,' and 'rejection' of transplanted tissues and organs;

> the capacity of the body of a recipient to 'recognize' tissues and organs that are 'foreign' to it, and to distinguish 'self' from 'nonself' […]and the 'chimeras' of genetically different groups of donor and recipient cells that are formed. (236)

The gift of the organ transplant comes either at the cost of impairing another individual's organism, or after one's demise. Through this brief scene, Almodóvar questions the authenticity of the human body. Could a body that has received an organ transplant still be considered unmistakably personal? For the Spanish director, one cannot be authentic unless one is simultaneously chimerical and tolerant. Consequently, there is no universality in authenticity (and probably no universality whatsoever). Authenticity is another way of individualization and self-acceptance, but again, if we have difficulties accepting the others' uniqueness, a perfect example could be taken from the chemical language that exists between the *non-human* entities of aphids and plants: "[i]n response to an overabundance of aphids on their leaves, [they] call ladybugs to their rescue by means of a language of chemical scents. Such behavior turns out to be but one of the many ways in which plants make their mark on the world, and it suggests that even plants possess a kind of agency" (Bennett 170). This film's tribute to presence/present is a necessary reaction to adapting the myth of authenticity, which, bluntly put it, is all about being true to oneself.

Furthermore, this film is probably his greatest achievement in the trilogy dedicated to delicate situations and/or medical conditions (the two other films are *Women on a Verge of a Nervous Breakdown* and *Talk to Her*). In all three, he puts us in an extreme situation where we have to observe characters having bouts of nervous breakdowns (with no much hopeful remedy), being in a coma (where the caregiver is completely eliminated at the end) or debating over the ethical aspects of organ transplants. In *All about My Mother*, it seems that he wants to re-write the "laws" established underneath xenophobia—the compulsive fear of the foreigner—by using a very subtle analogy of the organ transplant. When facing a life-or-death situation, a patient does not have time to question the identity or the status of the donor because *all* s/he wants is to be saved. Even more poignantly, I contend, he places this xenophobia in a medical context of the stillborn child and of adults in persistent vegetative state. More explicitly, Elizabeth Hallan, Jenny Hockey and Glennys Howarth believe that in those two unfortunate cases, the

> [s]ocial identity becomes unclear in the sense that the binary relationship of differentiation between 'life' and 'death' has been destabilized. […] More recent practices such as cuddling, dressing, naming, photographing

> and providing a funeral for the stillborn body now invest it with a social identity as 'daughter' or 'son.' Conversely, the brain dead body, which has a 'life-like' appearance, displays signs which effectively sustain a pre-existing set of social relations. For relatives, therefore, the brain dead individual's previous social identity persists into the present. (71-2)

Because cinema has been having an enormous impact on people of different backgrounds and education, Almodóvar's film intends to make us aware of our unjustified fear of the other, as well as of our obsession to secure our body's authenticity. The organ transplant, even though is presented only briefly, represents in fact the main concern proposed by this film. If a body in crisis accepts an organ transplant after minimal, if at all, checkup, then shouldn't we learn how to moderate our xenophobia? Also, through the organ transplant scene, Almodóvar informs us that the demised body's internal architecture succumbs gradually to death; or, as Lupton notes, "What is the status of a person's organs which have been taken from a body pronounced to be dead and transplanted into a living person's body? [...] In medical discourses relating to the viability of the human body organs and tissue for transplantation or other medical use, tissue may be described as 'dead,' 'double dead' or 'triple dead'" (*Medicine* 51).

Drawing the moral conclusions of this film as it has been proposed by the anonymous patient's body's acceptance of the organ transplant, in subtext, Almodóvar reminds us that it is the mind (echoing the institutionalized discourses)—and not the body—that has created extreme notions such as "abnormal," "teras," "xenos," "abject," etc. In return, the body is tested to see if it accepts/rejects these notions, copying at a smaller scale the process involved in an organ transplant. Finally, the organ transplant, which could be also decoded as the metaphor of liminality (placing the body at the threshold between life or death), is a reminder that modern medicine has a long time ago moved away from Galenic anatomical practices, where the body was generally left intact, considering it to be a sacred entity. On the other hand, modern medicine practices autopsies in its attempt to examine more closely the affected tissues and suggest remedies for future patients, as well as try organ transplants to save the lives of living persons.

Almodóvar's tribute to (non-)authenticity may finally be seen as a possible answer to our dream of immortality. According to this artist, the dream of immortality could mean going out of one's original embodiment to live in another human being's body via organ transplant. By so doing, it appears that the dream of immortality means losing oneself in an embrace with the unpredictable other.

As inferred from the analysis of Almodóvar's film, the two canonical concepts of "self" and "authenticity" would need to undergo a semantic transformation to fit into our contemporary definition of being. One's self's authenticity is a repository of one's desires and frustrations as retrieved from the inextinguishable well of the collective unconscious as well as institutionalized discourses. One's self becomes authentic if one loses it several times during one's lifetime, and thus periodically allowing it to reinvent itself. Therefore, one's self is never truly anchored in its essence; instead, it embraces its manifold manifestations of its mini-episodes of identity-seeking. In other words, we are engaged into a process of adding and subtracting moments to and from our lives, each with its presumed uniqueness, without realizing the repeatability of our gestures, the recurrence of our habits, and the desire to authenticate and give a precise meaning (definition) to our relationships and identity.

Concluding Remarks

As stated in the beginning of this chapter, during sleep we experience the D (desynchronized) phase. Coincidentally, D also brings to mind Desire, unlimited and unfulfilled, Uroboros-like Desire. Our inner bodily sensations—the proprioception dimension of our sentient being— have their unique way of perceiving time and space as they are guided by stimuli from within. To express our ideas, we rely heavily on metaphorical constructions. By the same token, because our bodies are the fluid metaphors par excellence, discovering their final referent/embodiment is similar to our perception of the horizon, whose lips are constantly misleading and receding. The body experiences an effervescent folding, pouring itself into the future while keeping/storing the echoed versions of its former embodiments.

Then, perhaps, Horne's designs of ready-to-wear bodily accessories, Gonzales-Torres's approach to a body disseminating into multiplied copies, and Almodóvar's tribute to (non-)authenticity may not seem as heretical as they must have appeared in the beginning of this chapter. The body wants to leave something ahead of as well as behind itself. After we die, the chrysalis of the body may appear deserted, but the butterfly, whose wings touched uncountable objects and people during its lifetime, has finally flown higher than its material coop. Immateriality may be the ultimate desire to the quest in finding our absolute presence. Lastly, the dichotomy presented in the title of this chapter requires a concluding remark here. We are born in a material body, which we *lose* in death (le corps perdu); however, our body also contains an immaterial thread that becomes the fabric of other people's memories, and thus continues us posthumously (le corps continué). Consequently, the immaterial body levitates because it has finally lost its burdensome weight and is now free to disobey the laws of physics.

Notes

[1] In 2005, Natalie Horne received her MFA degree in Art and Design from Purdue University. Currently, she works as an Art Rental Officer for the Saskatchewan Arts Board, in Regina, Canada.

[2] Is it ethical to transluminate one's dead body? Since this question is too sensitive to deserve only a parenthetical note, readers should consider that in 2008, the Nobel Prize in science was awarded to a group of three scientists who have been working on developing and improving the quality of the green jellyfish protein that could illuminate our inside (at the cell's level) and locate more precisely its malfunctioning.

[3] According to Jacques Lacan, during infancy, children perceive their body as fragmented parts and images. As adults, however, once we have established our distinct identity, we rarely pay attention to its fluid, changing essence. Horne believes we should regress to a stage in our lives where, this time consciously, to discover our bodies as being constituted from "pieces."

[4] The pamphlet is inserted below. Paragraphs two, three, six and seven contain a blank that should read "immortalis."

[5] The old idea of a genius who used to achieve eternity through his work is thus sarcastically rewritten.

Our Products . . .

Health - it's the newest accessory! Show off your vitality with custom-made, limited edition replacement organs, prosthesis parts, and multi-treatment pharmaceuticals. Allow yourself to experience life forever - in style!

Our Services ...

Never have the need to set foot inside of a hospital! All services provided by   may be performed either on site in one of our calm, luxurious treatment suites, or in the comfort of your very own home.   is also pleased to announce our latest concept - teletransport services! Enjoy convenience, privacy and relaxation at any of our various international locations via teletransport.

Our Guarantee . . .

Should you feel that your experience with   was unsatisfactory, or that you did not achieve the desired result,   will evaluate your concerns and compensate you accordingly. In the unlikely event of death,   will provide a full refund - pending the results of a post mortem analysis performed by our own team of pathologists. We are committed to maintaining full confidentiality of our clients records and refrain from disclosing information to any third party associates. Full-time 24-hour security is provided, ensuring that your privacy, comfort and safety is our #1 priority.

The world we live in today is rife with change, fluid in nature. Faced with ever-changing technologies, and endless choices of products and services, we are vulnerable to the lure of the quick fix and the designer seal of approval. The realm of consumerism has permeated nearly every aspect of our lives, forming an objectified material aura around all that we buy. My art is an exploration of the dichotomous relationship between needs and desires.

According to the 2004 National Center for Health Statistics, the average life expectancy of the American male is 74.5 years, while females can expect to live to the age of 79.9. Consider the idea of never growing old or aging, but instead living forever and enjoying all that life has to offer. Never worry about not having enough time to work, or more importantly, time to play! Everyday, advances in medical technology are making it possible to live a longer, fuller life. If you had the opportunity, wouldn't you want to life forever?

             , an exclusive and independent life company has undertaken the challenge of stopping the aging process and developing procedures to engage life into infinity.   replacement organs could be the answer to organ donor shortages. The United Network for Organ Sharing currently indicates that nearly 90 000 people are in need of transplants, while only approximately 8 500 donors are available. In addition to donor shortages, the current Census Bureau report that nearly 18.9% of Americans were without health insurance in 2004. Would you be willing, or able, to pay the price for   products and services?

The concept of   derives from a fascination with material culture and a desire for commodity goods and services. The most common exchange is of things in return for money. Even where health and human life are involved, the transaction of "things" is still prevalent. Death and illness are a part of what makes us aware of living. Simply maintaining or being grateful for life we have is a challenge for many, yet for some it will never be enough.

- NRH 2005

New York - Los Angeles - Palm Beach
- Everywhere -

Phone : 1 800 Immortal

immortalis:

(noun) Not mortal; not liable or subject to death; deathless, undying; living forever.

Oxford English Dictionary

Our Company . . .

immortalis™ is an exclusive company, a brand, and a way of life envisaged by one person's simple desire to live in excellent health, forever. By filling a niche market, catering to only the most discriminating individuals, ***immortalis™*** redefines how healthcare is conceived, utilized, and delivered. Through the collaborative efforts of health professionals, scientests and designers - vitality has a whole new look, meaning, and experience.

Our Philosophy . . .

At ***immortalis™*** we strive to offer the best in quality of care and total - life products. Our elite standards are set by your demands, and we continue to aim beyond those standards. To provide you with the full, complete luxury life package is our ultimate goal. As our motto says:

Luxe Life . . . ∞

Product & Service Information

1. Replacement Organs - Custom-made out of only the finest biomedical plastics and adorned with Swarovski crystals. Become a whole new person with our replacement organs.

Heart	$35 000.00
Lungs	$32 500.00
Kidney	$28 000.00
Liver	$25 000.00
Trachea	$28 000.00
Uterus	$30 000.00

2. Prosthesis Parts - Professionally sculpted prosthesis parts, made to look like the real thing - only better! Make an appointment for a personal consultation. Prosthesis include but are not limited to:

Rhino-replacement	$5 000.00
Maxillofacial-replacement	$6 500.00
Umbilico-replacement	$3 000.00
Oto-replacement	$4 000.00
Carpal-replacement	$8 000.00
Tarsal-replacement	$8 000.00

3. Orthopaedic Reinforcements & Joint Replacements -Regain the agility of youth with ***immortalis's™*** specialty orthopaedic reinforcements and joint replacements. Your body will thank you!

Femoral-reinforcement (nail)	**$10 000.00**
Humurous-reinforcement (plate)	**$8 000.00**
Joint-replacement (hip)	**$25 000.00**
(knee)	**$28 000.00**
(shoulder)	**$22 000.00**

4. Internal Physical Imaging - Determine the quality of your health with immortalis's™ state of the art diagnostic imaging services. You will believe the results when you **see** them!

X-Rays	$800.00
Computer Axial Tomography	$1 000.00
Magnetic Resonance Imaging	$1 500.00

5. Multi-treatment Pharmaceuticals - Why take several pills to cure ailments when just one can do the trick? Make an appointment for a consultion with ***Immortalis's™*** own pharmacists and chemists, who can custom formulate that "one pill" to treat everything!

Based on a 30 day average	$3 000.00

* Price will vary depending on the conditions being treated and the duration of the prescription.

6. Additional time in recovery suites - Take your time recovering from procedures in any of our fully furnished deluxe suites. Inquire about our special ***Rejuvination Chaise***, designed to provide elevation, comfort, and relaxation!

One Night Stay	$1 500.00
Three Night Stay	$4 000.00
One Week Stay	$10 000.00

Accepted Forms of Payment . .

Payment for ***immortalis™*** products and services is by cash or approved checques only.

immortalis™ does not accept any form of health insurance.

*Inquire about our special VIP price benefits!

[6] This remark could become emblematic for his "giving-taking," or the double nature of his artistic gesture.

CHAPTER TWO

RECLAIMED BODY

I experience myself as embodied.
In the realm of medicine, the body
is rendered an object. It is inspected,
palpated, poked into, cut open. From
being a locus of self, the body is
transformed into an object of
scrutiny.[…] Medicine does not
articulate an etiquette of the body.
(Young, 1)

As persons become patients, they relinquish
their social personae. They divest themselves
of some of their social properties with their
clothes. […] The boundary of the self is
not ordinarily coterminous with the skin.
[…] Constraining the self to its
bodily integument is a move toward
rendering the body an object.
(Young, 14)

In her book, *Presence in the Flesh: The Body in Medicine* (1996), Katherine Young tackles the issue of the body as object. She perceives the body as continuing beyond its physical territory, extending spatially to surrounding objects with which it has been in contact. Through the act of touching, the body enters into a dialogue with these objects. The more frequently performed these dialogues, the better the chance for a body to form its hexis, "[f]rom the Greek word habit" (83), thus enveloping itself in a space of secured familiarity. On the other hand, Young argues that, under medical examination, the body loses its subjectivity by becoming an object of analysis and control. In this light, medicine is viewed as being concerned not with the name of a patient, but with the name of his illness. Or, we could say that medicine is interested in going beyond a body's subjectivity to find out what caused its breakage into illness as a potential way of restoring its wholeness.

There are other ways to think of our bodies as objects. One example, also presented in Young's book, comes via Drew Leder's notion of bodily withdrawal. According to Young, our internal organs—not visible to us—participate in a series of ecstatic processes. Though we cannot see them, their internal activity generates the shaping and reshaping of our bodies' architecture. As Young acknowledges, "My interior is radically inaccessible to me. I am hard put to influence by conscious intent, my digestion, or my heartbeat, or my brain waves. […] It is this withdrawal of the viscera […] that permits the objectification of the body" (87).

In addition to examining the ways of seeing the body as an object, it is important to mention the changes that intervene in the sedated body. In such cases, the body becomes an object transformed not only by an illness, but also, and even more dramatically, by drugs that tranquilize a patient. When the body is rendered too weak even to be aware of its physicality, it becomes an *unconscious* object. What happens in those instances when there is little or no hope at all for patient recovery? How could these patients reclaim their bodies, and their personhood?

In this chapter, I argue that medical equipment and/or strong tranquilizers do not reinforce one person's individuality; on the contrary, they make patients feel like medical cases. Drawing some of my argument from Young's idea that "constraining the self to its bodily integument," as medicine does, "is a move toward rendering the body an object" 14), I investigate how a patient's visceral semantics, while in pain, denies this impersonal vision of a body qua object, thus seeking ways to repossess its (temporarily) lost subjectivity.

Ne Habeas Corpus

> These puppets have the advantage of being antigravitational. They know nothing of the inertia of matter. [...] These puppets, like elves, need the ground only to brush it. [...] We need it to rest... (Kleist "On the Marionette Theater," qtd. in Caruth, 73)

The above quotation appears in Cathy Caruth's book *Unclaimed Experience: Trauma, Narrative, and History* (1996). The German writer Heinrich von Kleist's analogy between puppets and people emphasizes the difference between mere movement and refined kinetic abilities. Kleist's analogy provides insight into the experiences of two paralyzed men: Ramón (of Alejandro Amenábar's 2004 film *The Sea Inside*), who snapped his neck against the sea floor and Ken (of Brian Clark's 1979 play *Whose Life Is It Anyway?*), who was crippled in a terrible car accident. As a consequence of their misfortunes, they will never be able to walk again. Although these two men exhibit an undeniable drive toward Thanatos, there is nothing Freudian in it because it is not linked to or complemented by Eros.

The main argument of this chapter's section is constructed around the notion of *habeas corpus*, which was introduced into the legal system to protect the rights of prisoners who were being held by the state unlawfully. Under this concept, they were "presenting" their bodies to the court, which then had to make a determination of whether the person should be held in custody or released. While in Latin "You may have the body" is a jussive construction, I add a "*Ne*" to reflect upon those instances when the body, although still *de facto* personal, cannot participate actively in life.

Prior to their life-changing accidents, Ramón and Ken could function normally in society. Now, paralyzed from neck down, the two characters tragically illustrate the Cartesian split between mind and body. I find it pertinent to redefine the concept of the "bodymind" as it has been proposed by theorists such as John Dewey, Antonio Damasio, and Floyd Merrell. The bodymind is the enchanted space where our mind touches its corporeal epidermis and beyond. Or, as Merrell argues in his book *Sensing Corporeally: Toward a Posthuman Understanding* (2003), "[b]odymind or bodymindsigns [...] [are] operating in concert with one another to the extent that there is no knowing where one leaves off and another begins" (161). Needless to say, for Ken and Ramón the bodymind is an unusable, static concept. Even more dramatically, they refuse to live a stationary life in which they believe there is no dignity anymore (Ramón speaks of "RDD," an acronym for "Right to Die with Dignity").

In addition, the men's condition represents another ontological challenge: Since there is no hope of restoring their previous *status quo*, they defy, though unintentionally, the condition of being a patient. This is a process with many meanings that increase in intensity during the period of convalescence. The experience is also a perverse test of our ability to accept and cope with unexpected events in our lives. As we traverse insecure places throughout our lives, we want to define what we feel as clearly as possible. Physicist Niels Bohr has influenced the fields of literature and medicine through his works and scientific discoveries. He coined "the principle of complementarity," namely those phenomena when something possesses two different natures or essences simultaneously. A classic example in the area of physics, the electron acts both as a particle and a wave at the same time. Bohr also said that in this life we are both spectators and actors. Comparatively, the field of medicine and literature has advanced the theory according to which health and sickness should be regarded together, as an inseparable whole. This naturally poses the question, Why do we think of ourselves as sick *only* when are sick, and as patients *only* once we become patients? Where is the "and," the vital copula in those critical moments? Usually, a person experiencing severe pain becomes a patient; then his suffering becomes a medical case (as recorded and transcribed in medical charts).

Throughout this section, I refer to Ramón and Ken neither as "persons" nor as "patients," but as persons/patients, because once someone becomes a patient he cannot fully return to being the person he was. One cannot forget one's experiences as a patient. In *AIDS and Its Metaphors* (1989), Sontag reminds us that "etymologically, patient means sufferer. [Nonetheless,] it is not suffering as such that is most deeply feared but suffering that degrades" (37). A patient straddles a line, during which pain pushes him into assuming a different identity. Julia Epstein argues that "A patient's history translates the bodily aspects of a life into a documented aggregate of symptoms and turns what might be a human story into the professional discourse of medicine. […] Only by this process of converting patient into case can [the medical system] bring to bear its tools of interpretation and therapeutics" (55).

I propose two more interpretations for being a person/patient. The first one reads "patient" as an adjective, emphasizing the struggle to effectively learn to adjust to a new physical and psychical *status quo*. The second, when the indefinite article "a" is not put in parenthesis, it illustrates that most patients realize the transition that has been made in their lives; not only are they now *a* patient seeking help, but they are also *a* medical case. Consequently, they have a difficult time accepting their temporarily confused identity.

While many patients retain a hope of being restored to their previous condition, unfortunately, there exists none in either Ken's or Ramón's case. Without hope of recovery, a crucial element is missing from the curative ritual, and they lose their ability to feel like human beings because they cannot enjoy the synchronized encounters between their mind and body. In their case, that encounter is forever lost. They do not possess/have a bodymind anymore, but a body that is sketched as well as secluded at the mind's level. As Ken admits: Look at me. I can do nothing, not even the basic primitive functions. I cannot even urinate. […] Only my brain functions unimpaired but even that is futile because I cannot act on my conclusions it comes to (142).

There is an unfortunate, yet illustrative, breakdown between his mind and body. Although his body is supplied with fluids, glucose, food, and pills, these aids fail to recreate the link between mind and body. The great irony is that, to a certain extent, Ken's mind has remained unaltered during the complete change of his body.

This state of mind's non-alterity versus the body's severe deterioration is completely different than the typically known mind/body dualism. As argued in her book, *The Flight to Objectivity: Essays on Cartasianism and Culture* (1997), Bordo points out that:

> In Descartes, [...] a crucial departure was made from the Aristotelian and medieval epistemologies: from the notion of 'mind-as-reason' to the notion of 'mind-as-consciousness.' For Aristotle, there had been two modes of knowing, corresponding to the two ways that things may be known. On the one hand, there is *sensing*, which is the province of the body, and which is the particular and material; on the other hand, there is *thought* (or reason), which is of the universal and immaterial. For Descartes, both these modes of knowing become subsumed under the category of *penser*, which embraces perceptions, images, ideas, pains, and volitions alike. [...] The characteristic that unites all these states is that they are all *conscious* states. (50)

Moreover, Ken's and Ramón's *penser* is overwhelmed by an excess of thinking. But "to think" is not enough to justify their existence; if anything, to think becomes a burden for them. Because their bodies are deprived of their innate capacity to move, the men cannot perceive the variations that require physical action. According to Julia Lawton, "patients' conceptions of time (and, therefore, of themselves in time) had become radically different than that of their family and friends" (46-47). For Ken and Ramón, time is not entropic anymore, but self-contained--full with memories that are agonizing and do not offer any comfort. Merleau-Ponty believes that "existence is spatial, [namely] that through an inner necessity it opens on to an 'outside,' so that one can speak of a mental space" (351).

However, when living exclusively *in* a "mental space," Ken and Ramón realize that, while time is motionless for them, it gallops "out there," for those who can walk. Paralleling the space-time continuum, I propose the creation of a perception-action continuum; in this particular case, however, Ken and Ramón experience a mind-body *discontinuum*, no matter how bizarre this may sound to us. In other words, they experience a state of frustrating imbalance. Their bodies are lifeless and their minds are active to the point of being neurotic. For them, action becomes thought; or, even more precisely, movement becomes a series of cacophonic, loud, recurrent in theme—yet *non-sequitur* in action--thoughts.

According to Bordo, "To be able to mentally represent an object in its absence is to conceive of the object as constituted not by this or that transitory perception of it by the subject, but as sustained by a projected multiplicity of perspectives—as having a 'being-for-others'" (46). In the case of those severely injured, the body becomes an absent object, both for themselves and for others. In phenomenological parlance, these persons/patients do not know anymore what it feels like to have a body versus to be a body. Their bodies are a massive lump always sitting inertly in bed. Disconnected from the body, their minds are engulfed by one central thought: suicide. As Ramón acknowledges, "[a] life that ends liberty, is not life at all." He writes his memoirs, "*Letters from Hell*," to serve as a testimony of his ordeal and an appeal of *ne habeas corpus*. Not having an active body, he cannot find comfort in the notion of *habeas corpus*. What he may argue is that *Solam meam mentem nunc habeo* (i.e., "Now I have only my mind.").

But that is not hope in his Pandora's box; instead, it is his curse. Before he starts videotaping his death, he admits that "living is a right, not an obligation, as it has been in my case." Artistically, it has always been difficult to bring about closure in a film whose main character's ordeal has not ended. In *The Sea Inside*, Ramón commits suicide, but what is really traumatic is the orchestrated synchronicity of his suicidal act with our watching of it.

For the moment, it is sufficient to emphasize that Ken and Ramón both experience a loss of contact with their own bodies and other bodies. Merleau-Ponty in his phenomenological philosophy creates the theory of a body as enveloped spaces (i.e., the body in its multiple embodiments: lived body, living body, and having a body). In a different field, Mikhail Bakhtin speaks about dialogism and heteroglossia, and how one's utterances acquire more depth and value as they become incorporated into another person's dialogue. In other words, the dialogic dimension of the self is reciprocated to a certain extent by the dialogical body. The best way to describe the dialogical body is by saying that it bears the effects of the body multiplied in discourses and actions. The bodies of Ken and Ramón, however, offer a radically different example. A relevant moment occurs when the social worker visits Ken in the hospital; advising him to start reeducating his mind by developing an "occupational therapy" (68), the latter responds sarcastically:

> KEN. I might even learn to do wonderful things, like turn the pages of a book with some miracle of modern science, or to type letters with flicking my eyelids. […] I do not want to become happy by becoming the computer section of a complex machine. And morally, you must accept my decision [to die]. (69)

With their bodies turned more and more into superfluous, evanescent thoughts, Ken and Ramón are left with only a memory of their former embodiments, which they want to tear apart. They are the perfect example of the *sparagmos* motif. According to the legend, Dionysus' limbs were dismembered by the anger of the Titans. But since Dionysus was a son of Zeus, hence immortal, his members were not permanently affected by this brutal tearing, and were made whole again. In Ken's and Ramón's case, however, their bodies—once broken—will stay broken. It goes without saying that we experience ourselves seriatim; we have a more or less developed awareness of our corporeality. On the other hand, these two men experience themselves locally, in their mind. Is pain in such cases psychogenic?

Before considering this controversial question, it is useful to note that psychogenic refers to those disordered states that are mental/emotional rather than both mental and physical/physiological. Even more poignantly, based on the experience of Ken and Ramón, I interpret psychogenic functioning as a metaphor of imprisonment; consequently, for me psychogenic refers to a person whose corporeality is literally locked in his mind.

To understand the importance of imprisonment, it is useful to make a digression. In his book, *Discipline and Punish: The Birth of the Prison* (1977), Michel Foucault notes, "[p]rison continues a work begun elsewhere, which the whole of society pursues on each individual through innumerable mechanisms of discipline" (303). What are these "mechanisms of discipline"? One person's discipline starts during the early years of one's life, in one's own family. It continues in schools and later at work. Sometimes, it even continues in hospitals, prisons, and monasteries. No matter where one's discipline starts and where it ends, we all have experienced its dramatic consequences: lack of energy, moments of immense doubt about our role in society, times when we suffer from ennui, *déjà vu*, and routine—all of these being nothing more than partial synonyms of incarceration and direct or indirect inflicted punishment. As Foucault argues further, "In its function, the power to punish is not essentially different from that of curing or educating" (303).

The power to punish, as well as that to cure and educate, all depend upon coercive words and persuasive mechanisms. Are bodies in (severe) pain coerced to discipline, and could this be one reason why they are sedated? If so, could we say they inevitably become docile bodies? In her book, *The Dying Process: Patients' Experiences of Palliative Care* (2000), Julia Lawton argues:

> In removing a patient's sentience through sedation the last vestiges of their personhood are also erased. [...] It thus appears that the practice of sedation is not grounded solely in an ethics of beneficence. [...] It actually requires little imagination to argue that sedation was in fact performed primarily to reinforce the hospice's ideology of a good death, particularly in view [...] that such an ideology requires the location of 'docile bodies' within communal space. (120-21)

Ken and Ramón have become persons/patients with inarticulate bodies, which--helped by machines--could live indefinitely. As seen in this play and film, joy and anguish have been replaced by medical equipment and strong tranquilizers that keep these two men alive in spite of their desire to die. The wound of their bodies is now exclusively the wound of their minds, since the latter is their only percipient site. According to Caruth, "[t]he Greek word *trauma*, or 'wound,' originally referring to an injury inflicted on the body. In its larger usage, […] is understood as a wound inflicted not upon the body but upon the mind" (3).

For Ramón and Ken, distance is a paradoxical concept. They feel distant from their former selves/identities because nothing that had existed in their past can define who are they now; on the other hand, they cannot see themselves within this distance because they are literally locked in their past. Their desire to die creates a thought-provoking debate on the controversial aspects of euthanasia and personal rights. Anne Hunsaker Hawkins suggestively notes that "[t]he tendency in contemporary medical practice is to focus primarily not on the needs of the individual who is sick but on the nomothetic condition that we call disease" (6). Once a body is diagnosed and becomes a "case," does it belong to the person, or to society? Medicine, law, religion and other institutionalized and marketable sites of power try to keep humans together, as a group, as a *flock* of individuals, putting us into certain categorizes. To the medical world, we are either healthy or sick; to the law we are either good citizens or wrong-doers; and to religion we are either moral or immoral. Other dichotomies could be created easily, thus perpetuating an ineffectual (Cartesian) way of reasoning. With no hope whatsoever and degraded by suffering, Ramón and Ken have pointed out that the body is a social product and construct. However, as persons/patients in pain they would appear to have the right to reclaim their body, thus attributing to physicality its supreme value over its socio-cultural readings/functions/roles.

Both the play and the film emphasize the discrepancy or conflict that exists between two fundamental institutions--the law and the medical field--where each try to impose its hierarchical supremacy over the other. As Ken confesses to Hill, his solicitor who will represent his plead of being allowed to die: "I'm almost completely paralyzed and I always will be. I shall never be discharged by the hospital. […] I therefore want to be discharged to die" (76). Later in the play, Hill explains to Ken how ironically problematic it is for him to reclaim his own body:

> HILL. Under the Mental Health Act of 1959, […] a doctor can keep [patients in the hospital] and give [them] what treatment he thinks fit. (85)

Needless to say, some treatments are either not fully explained to patients, or abusively prescribed to them (namely, without having the patients' prior consent). In this context, the following series of retorts between Ken and Dr. Emerson is emblematic:

> KEN. Doctor, I did not give you permission to stick a needle in me. Why did you do it?
>
> DR. EMERSON. It was necessary. You will find that as you gain acceptance of the situation and you will be able to find a new way of living. (44)

From the very beginning of the play, Ken cannot tolerate the hypocrisy that pervades medical practices and discourses, and how much this affects the morale of a patient. As he confesses, "Everyone who deals with me acts as though, for the first time in the history of medical science, a ruptured spinal column will heal itself" (5).

Ramón is not a stranger of this generated pretense either. Actually, his case seems to be even more dramatic, for he has spent twenty-six years confined in bed. With half a lifetime spent in bed, he has been absent from many of its joys and frustrations. According to Leder, "The lived body, as ecstatic in nature, is that which is away from itself. Yet this absence is not equivalent to a simple void, a mere lack of being. The notion of being is after all present in the very word of absence" (22). We may argue further that just as we do not have one body but several embodiments, comparatively we experience several absences (some productive, while others are destructive) from our being-in-us and "being-in-the-world."

When Teresa de Lauretis writes "Experience is never *im*-mediately accessible" (qtd. in Haraway 142) what she may be implying is that a person constitutes himself as subject if and only if his subjectivity becomes temporarily blocked by objects to be contemplated and events to be intrigued by. Otherwise, without this suspension from one's experience, there would be an unproductive and disturbing continuum of ego-subject-subjectivity. An experience that is never "im-mediately" accessible may also mean that there is a necessary rupture not only between the subject immersed into seeing and reflecting upon an object, and sometimes upon himself as an object, but also that our lives consist of inescapable moments of partial absences.

One classical example of partial absence is experienced by all of us during sleep, when we put our turmoil on hold by disseminating ourselves into dreams. According to Cataldi, "Night [...] envelopes me and infiltrates through all my senses, stifling my recollections and almost destroying my personal identity. [Night] is pure depth without foreground or background, without surfaces and without distance separating from me. [T]his 'pure' depth presents itself as a candidate for the 'absolute' space" (48). Sleep is a recurrent theme for Ken and Ramón. However, in their case, almost invariably sleep acquires a different meaning; these persons/ patients are sedated. Thus, sleep becomes a regular activity, and sometimes their only activity. Their bodies are absent from direct experience. Once their spinal cord is fractured, they descend into a space where, in order to reclaim their identities, they must reconcile with their liv*ed*, consum*ed* past, as well as accept the intervention (or intrusion) of highly sophisticated machines that keep their bodies alive.

After the shock of their accident, they realize that they experience the regressive dimension of time, where they sojourn mentally in their memories. Having atrophied bodies, they traverse and then, although unintentionally, remodel the original version of the phenomenological chiasm—an otherwise necessary gap or space that opens between perceiver and perceived, subject and object, and myself and others. There is, according to Merleau-Ponty, an *écart* (or "divergence") that keeps one divided into subject and object. In the case of Ken and Ramón, this *écart* has been completely annihilated, since there is no genuine interaction between their bodies and minds. Essentially, their bodies have morphed into objects. Furthermore, they sculpt out their body from remembered sensations: how waiting for something once felt, how walking felt, or how pretty much *life* felt. By so doing, they reclaim their lives in an attempt to anchor these felt sensations in a now broken, failing body.

In fact, this recollection gives rise to a new type of embodiment, which I call reembodiment. The body is recreated or made-up not through actions and reactions to events, but it is constructed through the act of remembering life, hence its synthetic or artificial attribute. As Young explains, "My interior is radically inaccessible to me. I am hard put to influence by conscious intent, my digestion, my heartbeat or my brainwaves. […] In is this withdrawal of the viscera, their inaccessibility to our introspective attention that permits the objectification of the body" (87). For Ken and Ramón, there is a concentric or double withdrawal. While for each of us, there is a necessary withdrawal of the viscera to allow their proper functioning, for these two paralyzed men, that withdrawal is not purposely felt. This is why their mind searches to anchor the body's former sensations. The body, an otherwise symbolic entity, in this case becomes the mind's literal creation, a tapestry consisting of interwoven memory-like fabrics. The body becomes bodymind*ed*, refracted and reflected by the mind's uninterrupted exercise to replicate a former felt physicality. The made-up body, the bodyminded, or the reembodied body are all synonyms of a person/patient who attempts to reclaim his individuality, and thus refuse to be considered another medical case.

Even though their hands are not capable to exercise their endowed prehensile qualities, or their feet do not stomp grounds anymore, there is yet one more aspect of this made-up body that needs to be addressed. Fifty years ago, the Intensive Care Units were at their very incipient phase. Nowadays, each of us could potentially reside for a short or long period of time in an I.C.U. As physician Atul Gawande remarks,

> Intensive-care units take artificial control of failing bodies. Typically, this involves a panoply of technology—a mechanical ventilator and perhaps a tracheostomy tube if the lungs have failed, an aortic balloon pump if the heart has given out, a dialysis machine if the kidneys don't work. When you are unconscious and can't eat, silicone tubing can be surgically inserted into the stomach or intestine for formula feeding. If the intestines are too damaged, solutions of amino acids, fatty acids, and glucose can be infused directly into the bloodstream. (85)

If these two persons/patients live, it is precisely because of the advanced technological revolutions undergone within the medical field. Point in fact, "inert," an adjective that used to have negative and scary meanings, could today exhibit its specialized or modified connotations. Without doubt, Ken's and Ramón's bodies are inert, but at the same time, they are still alive. In other words, their minds would not have been capable to retrieve former, felt sensations, if the body was *indeed* inert, or destroyed. Paralysis has now new meanings, or as Dr. Emerson admits, "a doctor cannot accept the choice for death; he's committed to life" (Clark 91). This is a very disturbing paradox or fact; while these two men cannot move, their bodies are not exactly dead either. They need intravenous food to keep their organs working. Moreover, their bodies still leak and emit smells, which is another indication of their sustained internal activity.

In this chapter's section we meet two persons/patients in two different circumstances, but who share something important. According to Young, "as persons become patients, they relinquish their social personae. They divest themselves of some of their social properties with their clothes. Taking off layers of clothing circumscribes the self by limiting its extensions into social space" (14). Continuing this idea, it is useful to note that Ramón's sister-in-law takes care of him, while Ken is in a hospital. One wears casual clothes, while the other wears hospital gowns. Yet they do not exactly relinquish their social identities when they take off layers of clothing. Something else is at stake here, namely how their bodies become and function like a cluster of memories.

Due to their severe and crippling accidents, Ken and Ramón have entered into the category of "medical cases" and are forced to face the difficult challenges brought on by their new embodiments. Having now scheduled appointments with their doctors, or being confined to their very uncomfortable Procustean-like beds, their intimate bodies became social bodies. As Young suggests, "What it is to be a person is a cultural notion, one of the epiphenomena of ideology" (87). But ideologies are not adequate discourses when any disequilibrium occurs at the level of society. Ideologies are manipulative and restrictive. They envision a certain type of individual to fit into their norms, and have little to say/offer when people fall ill or experience traumatic events. Understood as such, an individual is an epiphenomenon consisting of those events that are not exclusively presented in ideologies, just as comparatively, there are iatrogenic complications resulting from a prescribed treatment to which a patient's body needs to adapt.

Furthermore, as Young argues, "patients fail to produce at least two necessarily sequenced clauses of which the second is consequential on the first. [...] Patients produce 'replays' that do not achieve the status of fully-fledged narratives, stories that do not come to an end" (69). This does not imply that patients do not want to speak; instead, it means that they must first absorb an otherwise obtrusive medical jargon, and then investigate their illnesses or medical conditions. In an attempt to reclaim their individuality, both Ken and Ramón have expressed their thoughts and concerns, thus proving that they can produce more than "replays." In fact, they question the meaning of life as quadriplegics, more specifically, whether, after their accidents, they can still live for themselves or for others. More specifically, their family members do not accept euthanasia as a solution to end the two men's ordeal, while their doctors do not accept defeat, and, thus grant their patients' wish to die. (Even though, ironically, Ken and Ramón are given strong tranquilizers that keep them asleep and constantly pretty much inactive.)

Then, what/who represents the body in pain, the body that is barely functioning?

It is by now trivial to say that our bodies are tacit. They are tacit not because they cannot speak, but rather because their dialogues happen within, beyond the epidermis, in a realm of magnified perceptions. Sondra Perl develops these arguments in her book *Felt Sense: Writing with the Body* (2004), based on Eugene Gendlin's notion of the "felt sense," which he defines as calling "[a]ttention to what is just on the edge of our thinking but not yet articulated in words" (xiii). The felt sense is closely connected, if not derived from, Michel Polany's "tacit knowledge" of our minds and bodies; that is to say, the fact that we always know more than we are capable of transmitting to ourselves and others through words.

Another philosopher interested in this idea, Ludwig Wittgenstein, once remarked that our ineffable existence was related to the fact that words could not adequately represent our feelings and thoughts. Words, I suggest further, are partial or gliding signifiers of our varied embodied sentiments. On the other hand, many persons/patients may feel teased and terrified by the tacit dimension of their new embodiments. In other words, they may experience the frustrating, tacit dimension of the unknown *on* their bodies. They feel they have learned (but not necessarily mastered) new words, namely new medical terminologies. However, that, ironically, has not given them comfort in understanding, coping, and accepting their new condition. Having been face to face with an open lexicon, in which words may acquire different meanings each day, they may feel their bodies to be in a dangerous, chaotic phase. For example, "tired" could one day mean vomiting a lot, then the next day resting in bed motionless and/or apathetically. This type of body learns new words and/or new meanings to old words daily and the mind becomes overwhelmed in its effort to minutely catalogue each new term.

For Descartes, the body was always in one specific place, self-contained; thus, it had a *hic et nunc* quality: "I think, therefore I am," translated loosely into I think *right now*, therefore I am *right now*. In actuality, we know that we are confronted with and challenged by unknown or unrehearsed situations that have little to do with the self-contained aspect of our bodies. But when one cannot feel one's body, does this automatically suggest one loses one's capacity to think (and locate one's body)? The body has an implicit narrative embedded in its immune system, which prevents us from some injuries and complications. But Ken could actually die from an infection generated by his own body. As Kershaw, Ken's barrister, tells him, "I am informed that without a catheter the toxic substance will build up in your bloodstream and you will be slowly poisoned by your own blood" (Clark 115). The ill body gives rise to a story in which the traditional plot has been replaced by many unknown variables. The ill body may become an *êpidemos* (Greek for "foreigner")--namely that kind of body that may actually harm someone.

On the other hand, as Perl notes, "Felt sense is the physical place where we locate what the body knows" (4). The experience of Ken and Ramón shows us the opposite condition, when a body does NOT know. At that point, the body can become so overwhelmed by new sensations, symptoms, and reactions to treatments that it is not sure anymore of what it feels. Moreover, sometimes the body becomes too weak to link itself narratively with the mind. While the broken body is challenged to learn new words, these words do not reflect a personal choice, but come from medical textbooks.

Nonetheless, what could be more efficient in these cases may come through the non-discursive nature of our bodies. According to Richard Shusterman, the body

> [r]emains a promising place where discursive reason meets its limits, encounters its other, and can be given a therapeutic shock toward redirection. […] We share our bodies and bodily pleasures as much as we share our minds, as they are surely as public as our thoughts. […] To understand the body as the 'nondiscursive other,' we have to stop pushing words and start moving limbs. (128-29)

Shusterman's idea of bodily nondiscursiveness is connected directly to the much debated notion of freedom. The body is aware of a certain freedom that is not so easily attributed to the mind (erroneously, we tend to think of the mind as not being physical, when in reality it is). Shusterman writes, "We share our bodies and bodily pleasures as much as we share our minds, as they are surely as public as our thoughts" (129).

Even severely ill or injured bodies want to reclaim their identities, a fact never inscribed in the medical charts. Because the body wants to cut off the limitations located at the mind's level and because it wants to justify its purpose and identity, there is evidence that our bodies want to experience life other than linguistically or strictly ideologically. Both Ken and Ramón reflect upon the transitory nature of freedom. As Merleau-Ponty suggests, "To be born is both to be born of the world and to be born into the world. The world is already constituted, but also never completely constituted; in the first case, we are acted upon, in the second we are open to an infinite number of possibilities. […] There is, therefore, never determinism and never absolute choice" (527-28). But how could such a remark help us delineate between what is ill and what is healthy? Even more importantly, how much does an ill body challenge our notion of freedom?

In trying to find a definition to pain and suffering, we are fixated on keeping the illusion of our wholeness, as if without this definition we would be impaired and fragmented, incapable of functioning. The routinely asked question, "Where does it hurt?," reveals more than a dialogue between a physician and a person/patient. After all, there are many other people in different professions who (try to) define pain and suffering and realign them into some collective perspective on the human condition and what it *means* to be a living person. A physician, nurse, priest, therapist, friend or especially a stranger have different definitions of pain and suffering. We have actually created specialized therapists—algologists—whom we ask to mediate our (understanding of) pain. It appears that when in severe physical discomfort/breakage, we need to have as many perspectives on it and definitions as possible, to have others negotiate for us this delicate encounter.

So, where does it hurt? Before the era of painkillers, inconceivable for us today, pain was a test of one's body's endurance; pain was a punishment, it had to happen and it was eventually redeemable. However, we have evolved, so that today pain is a discomfort to be alleviated immediately and, in this respect, "Psychopharmacology [...] has succeeded in *reducing* our tolerance to the ups and downs of our emotional lives [...] by offering opportunities for a 'quick fix' solution to our 'problems'" (Williams, *Medicine* 147).

It is well documented in medical treatises that the more we think about pain the more we feel it because of our neuron assemblies. However, when a person/patient is potently drugged so as to not feel pain (like Ramón and Ken), not only does he stop feeling pain, but also ceases to value/enjoy his complete array of emotions. He becomes absorbed into that euphoria-like state, and hence liminal, not responding to many outside stimuli, though not being exactly dead either. Interestingly enough,

> The spatial ambiguity of the visceral depths is accentuated by the phenomenon of *referred pain*. A process taking place in one organ can experientially radiate to adjacent body areas or express itself in a distant location. [...] An almost magical transfer of experience is effected along both spatial and temporal dimensions, weaving the inner body into an ambiguous space. (Leder 41)

Thus, the question raised earlier, "Where does it hurt?," begins to trigger more complex answers (and even pose its own questions) than we may initially have thought it would.

To attempt to answer this question, let us return to Ramón. As a quadriplegic, he is isolated in his own room, far away from the sound and fury of the world; nonetheless, he still wants others to know about his ordeal. He writes his story with a stick placed in his mouth, employing gargantuan efforts--as one could only imagine. What he subtly tells us, his presumably healthy audience, is that we are afraid to lose what we have, our bodies. Even if we manage to never be diagnosed with a life-threatening illness, or experience a tragic accident, with the passing of years, our teeth lose their strength, our veins become like stones, and our vision, hearing and stamina start to be less efficient. We function at high risk and/or with damaged "components," whether or not we recognize this uncomfortable truth. When treatments are palliative, what type of healing occurs? We realize that healing is another word with an increasingly unsatisfactory meaning because it has too many interpretants. Like health, healing is a mediated word due to the interference of the "negotiators" of pain/health (a physician, a nurse, etc.) and the mystery involved in the body's internal ability and luck to heal and restore itself after a serious breakage.

The body has started to develop more and more diversified functions by adding made-up aides to reinforce its posthuman identity. Both Ken and Ramón have demonstrated how difficult it is to reclaim their own bodies once they become a medical case. However, as inferred from their examples, even healthy and/or not yet injured bodies may eventually be confronted physically to challenge the limits of the famous expression *habeas corpus*. To rephrase, since we have added auxiliary elements in order to maximize our performativity, who could still claim to have an irrefutably personal body? We can continue to believe that "we may have the body," but, when the body breaks down, we grasp its otherwise hidden social infrastructure.

Edson's *Wit*

If in the previous section, we met two men who would never be able to walk again, Margaret Edson's play *Wit* (1999) envisions a different type of imprisonment. In her play, the stage is empty, bare. Vivian Bearing--the protagonist--has advanced ovarian cancer. The emptiness of the stage parallels the baldness of the patient, as well as the simplicity of her hospital gowns. They are now parts of who she has become: a stranger to/in her own body. Vivian is completely dependent on machines (IV poles and others), and on chemically induced treatments that are the outside markers of her new identity. Her inside identity is challenged, too. It is important to note that, "In cancer, non-intelligent cells are multiplying, and you are being replaced by the non you" (Sontag, *Illness* 66). Can these "non-intelligent cells" completely replace/erase one person's/patient's identity?

Discussion of *Wit* puts into the spotlight the last days of a woman whose former individuality comes to her and us through flashbacks, and the protagonist's reactions to her cancer, hospitalization, and the loss of contact with the outside world. How does one perceive the encounter with a hospital's environment? According to Tom Chambers, "Ellipses usually occur [...] between periods of entrance into the medical setting. [...] The farther the character goes from the medical world, the greater the chance of ellipses" (179). In the case of Vivian, who is never going to leave the hospital, her ellipses occur only at the level of her mind when she recollects fragments of her former identity.

But who is Vivian? She is a professor who specializes in 17th century poetry, particularly in John Donne's metaphysical poems. Yet because of her cancer, she is now a student of illness. Furthermore, because her ovarian cancer is in an advanced stage, the doctors propose to her a very drastic treatment (some might want to add inhuman), about which they know little, if anything at all. As Vivian sadly admits, "Shrinking in metastatic tumors has not been documented" (37). Therefore, incapable of still having control over her body, unable to teach her students the beauty and difficulty of Donne's poems, Vivian performs one final role: that of a patient who has been isolated in a cold and mechanized environment, practically forgotten by everybody. If defined, what would be the pedagogy of medicine? How is it performed on our bodies? In order to approach possible answers, it helps to consider the relationship between patients and doctors.

Ronald A. Carson suggests that "The hyphen is a key to understand the relationship between patients and doctors. The hyphen simultaneously signifies separation and synergy, disjunction and conjunction" (171). Thus, "The hyphenated space in the doctor-patient relationship is a liminal place of ethical encounter, alternating voices and actions—back and forth, address and response" (Carson 180). However, in Edson's play, whenever doctors explain something to Vivian, her reactions appear printed side by side, an evocative case of the gap and/or miscommunication between her and Doctor Kelekian and his staff.

> DR. KELEKIAN. The antineoplastic will affect some healthy cells.
>
> VIVIAN.
> Antineoplastic. Anti: against. Neo: new. Plastic: to mold. Shaping.
> Antineoplastic. Against new Shaping. (9)

A diligent, passionate scholar, all her life Vivian has tried to understand the meaning of words. But "antineoplastic" is a cryptic hybrid for her. Even when she divides this word into its minimal segments of "anti," "neo," and "plastic," the individually analyzed prefixes and word, respectively, still do not cohere with one another. Because she cannot find a satisfactory meaning for this medical term, Vivian--the scholar--becomes more and more frustrated by her ignorance of medical terminology.

Furthermore, Vivian thinks she lives in a slippery body that does not allow her to define its present embodiment. Because of this insecurity, the playwright employs the flashback technique to express Vivian's personality. For example, in the recollection of the encounter between Vivian and her teacher, E.M. Ashford, the two women, debating one of John Donne's poems, say:

> E.M. ASHFORD. Nothing but a breath—a comma—separates life from life everlasting [...] Life, death. Soul, God. Past, present.
>
> VIVIAN. Life, death ... I see. It is a metaphysical conceit. It is wit!
>
> E.M. ASHFORD. It is *not wit*, Miss Bearing. It is truth.
>
> VIVIAN. The insuperable barrier between one thing and another is ... just a comma? (15).

"Life, death." Like Vivian, we do not realize that between the two there is "*just a comma*," an instant of breathing, an infinitesimal moment that potentially could transform life into death, existence into nonexistence. As Vivian painfully admits, "We are discussing life and death, and not in the abstract, either. We are discussing *my* life and *my* death [...] Now there is not time for metaphysical conceit, for wit...Now is a time for simplicity" (69). Thus, as she is approaching death slowly but inexorably, Vivian, the professor-scholar-mentor, seems to have lost any interest in sophisticated notions and ideas. Like the cancer which now has spread throughout her body, Vivian's language has hardened, becoming devoid of its former fluidity. As Elaine Scarry argues,

> To assent to words that through the thick agony of the body can be only dimly heard [...] is a way of saying [...] there is almost nothing left now, even this voice [i.e., of the one in abominable pain], the sounds I am making no longer form my words but the words of another. (35)

This moment could be understood as if the one who has been in pain has finally surrendered, not wanting anything anymore, not being capable of thinking beyond pain. It is as if Vivian has been expelled from her own locus of becoming, as if time has stopped and now pours itself violently, backward, inward, and centripetally toward an immobile center.

On her deathbed, Vivian says, "I am like a student and this is the final exam and I don't know what to put down because I don't understand the question and I am *running out of time*" (70). The unwritten, unspoken, yet understood meanings of Vivian's degenerating body raise the following questions: What is cancer? Does cancer have a meaning? Could we understand cancer other than through the physicality of the one in pain? Cancer almost makes language meaningless. Words--like cancerous cells that spread all over one's body--are eaten up by silences, interrupted by short sentences, and then followed by some more unbearable silences. I read Vivian's pain and her inevitable death through the signs inscribed on her body: she is bald, "has a central-venous-access catheter over her left breast, so that the IV tubing goes there, not in her arm" (4), vomits constantly, and has lost considerable weight.

> VIVIAN. In everything I have done, I have been steadfast, resolute—some would say in the extreme. Now, as you can see, I am distinguishing myself in illness. I have survived eight treatments of Hexamethophosphacil and Vinplain at the *full* dose… I have broken the record. I think Kelekian and Jason [his intern] foresee celebrity status for themselves upon the appearance of the journal article they will no doubt write about me. But I flatter myself. The article will not be about *me*, it will be about my ovaries. (53)

We could infer from this passage that the one in pain not only eventually succumbs to pain itself (when the body does not have resources to fight against the illness anymore), but also that the body of the ill person is unjustly claimed by the medical staff. After all, Kelekian and his interns, by writing an article about Vivian's ovaries, will not be considering Vivian as a whole body; the focus of their research has switched to her ovaries, thus treating Vivian metonymically. Therefore, could the metonymical body be considered a drastic variation upon the body as an object? According to Frank, "[t]he medical interpellation shifts its terms of hailing from disease to health, from the diagnostically fragmented patient to the new, medicalized, 'whole' patient" (46). As noted in the introduction of this book, Frank coined the "society in remission" concept; there, individuals, after they have been acquainted with the breakage in their bodies through severe pain, enter into a cycle or loop defined by relapses of pain. That is to say, they get reacquainted with the pain, either physically (when the body breaks again) or emotionally (through an enactment of the drama of pain), or sometimes both.

Therefore, while medicine tries to stabilize our bodies, ironically or not, our bodies constantly prove the endeavor futile. We are bodies within bodies, not one body, but multiple embodiments that change their morphology constantly accompanied by multifarious sensations and myriad reactions. To diagnose someone means to put that person's/patient's illness in one category. However, to treat him/her means to find a cure for his/her special case of illness. Sontag writes that "[c]ancer is not one but more than a hundred clinically distinct diseases, that each cancer has to be studied separately, and that what will eventually be developed is an array of cures, one for each of the different cancers" (*Illness* 59). Then, the role of the doctor is to attentively collect and interpret the information received about his/her person/patient. (This information comes to him/her through routine and/or sophisticated clinical tests, dialogues with the persons/ patients, and collaborations between a doctor and his/her staff.)

As has often been argued, doctors play the role of historians. But when events do not happen in the more or less remote past, but occur in the present, then, in a shift of reflexivity, the person/patient becomes the historian. Narrating his/her "present-past," the person/patient may do it succinctly or with abundant details, may be able to recollect in depth the events "as they happened," or may not be willing to share too much about his/her suddenly invaded intimacy. From history textbooks, we have learned that, if an event has at least two interpretations, then deciding which version is true becomes problematic. Put differently, when one's body is analyzed by the doctor-as-historian and by oneself-as-historian, then inevitably there occur conflicts in interpreting the events "as they happened."

These conflicts provide a pertinent example for showing that there are limitations in medicine and how it views its patients, just as there are limitations in any other science. In Vivian's case, having been diagnosed late in her life, the doctors knew that they would not be able to save her. Their mistake, however, was to treat Vivian just as a case, and not as a human being. As Leonardo Cassuto points out, "The case study relies on this continuing tension between the abstract (and general) and the concrete (and individual)" (123). But Vivian sadly admits, "Medical terms are less evocative [i.e., than John Donne's metaphysical conceits]. Still, I want to know what the doctors mean when they … anatomize me. […] My only defense is the acquisition of vocabulary" (44). Unlike her former fascinating acquisition of Donne's terms such as "ratiocination," "concatenation," and "coruscation" that have taken her a lifetime to savor with their multiple connotations, now Vivian feels not only that she is running out of time, but also that she is refused a genuine dialogue with her doctors. As she says, "In isolation, I am isolated. For once I can use a term literally" (47). She regrets not having been given the opportunity to communicate effectively with the medical staff, and, thus, of understanding more about the nature of her cancer.

She realizes as well that there is a more perverse dimension to isolation, since she must be isolated because her immune system is so low and defenseless that it may actually attack her body. When she is put in that isolated room, she is literally left alone with her body, which she finally sees more clearly. Yet it is not her body anymore, but something accompanied, surrounded and sustained by an orchestrated set of machines. Ironically, *she* justifies the existence and meaning of those machines. The machines could function without Vivian; they could be plugged and unplugged effortlessly, with a simple touch of a button. Sadly, it is Vivian who cannot function without them. This is the ultimate definition of isolation, which comes as a shock to her.

Only Susan, her primary nurse, is there to support and comfort Vivian. Their close relationship somehow resembles that of a midwife to a pregnant woman. The *only* difference is that Vivian is pregnant with death, and that in her role of "midwife" Susan teaches Vivian how less painfully to step over the liminal threshold between being and non-being, life and death. According to Carson, "The concept of liminality (from Latin *limen*, for 'threshold') refers to the ritual 'space' in which one is suspended between two worlds, neither here nor there, betwixt and between settled states of self, as in rites of passage, or, by extension, when experiencing illness" (180). The hyphenated relationship between doctors and patients is often replaced by the hyphenated relationship between nurses and patients, who approach them as persons/patients candidly by assisting them more effectively in their needs.

> SUSAN. Well, they [Kelekian and his staff] thought that drugs would make the tumor get smaller, and it has gotten a lot smaller. But the problem is that it started in new places too. They have learned a lot for their research […] There just is not good treatment for what you have yet, for advanced ovarian. I am sorry. They should have explained this. (67)

They should have, but they did not. Consequently, it is Susan who admits the truth to Vivian and who also explains the latter's choice of being resuscitated or not when her heart will stop. Being so focused on Vivian as a case, and not a person, the doctors forgot to discuss this most crucial option with Vivian. When her heart stops, Jason claims her body, and he calls for a code blue intervention by saying angrily to Susan, "She's Research!," to which Susan replies: "She's NO CODE!" (82). Ignoring Susan's intervention, Jason calls for a code blue team to resuscitate Vivian. Thus, the end of the play is constructed antithetically; on the one hand, the playwright presents the agitation, confusion, and lack of ethics on the part of the medical staff (with the exception of Susan) and, on the other hand, Vivian's entering into death with dignity, thus "claiming" her own body.

CODE BLUE TEAM.
(Reading) Do not resuscitate. Kelekian. Shit.
(the Code Team stops working)
JASON. Oh, God.
CODE TEAM.
- It is a doctor fuck-up.
- What is he [i.e., Jason], a resident?
- Got us here, called a code on a no-code.
JASON. Oh, God. (85)

AUTHOR.
She [Vivian] walks away from the scene, toward a little light. [...] The instant she is naked and beautiful, she reaches for the light ....

Through Vivian's nakedness, Edson may propose a way in which persons/patients could reclaim their personhood. Vivian enters, if not becomes, the invisible yet pervasive light. I imaginatively played Vivian's final part in my mind, and I have found sufficient resemblance between her gesture (i.e., that of tearing apart her hospital gowns) and Lucio Fontana's signature cuts in one of his many paintings gathered under the title *Concetti Spaziale* (i.e., "Spatial Concepts").

For Fontana, what lies in front of our eyes when we perceive a painting is only a metaphorical gate that leads us to experience more than one perception. Fontana was interested in finding something beyond the limitedness imposed by the canvas; he searched for an opening, and this is one of the reasons why he cut deep into his paintings. The resulting line could be compared to scars grown on the smooth texture of our skin; even more notably, visually, this vertical line invites us to see beyond it, to unveil what is not represented on the canvas, to assume there is something else besides what lies in front of our eyes, and thereby to incite our imagination. Furthermore, Fontana's paintings--united under the heading *Concetti Spaziale* -- were given an additional word, *Attese.* This word "[i]s usually translated as 'expectations' or 'waiting,' but whichever word one chooses, the sense of longing is always there" (Whitefield 29). Everything that exists in a state of becoming is never truly completed or transformed. These transformations are dependent upon the passing of time since they uninterruptedly record their successive manifestations.

Moreover, Fontana argued that "[t]he surface cannot be confined within the edges of the canvas, it extends into the surrounding space" (qtd. in Whitefield 136). Since we are living as well as lived organisms, extended and extending matter, we are accompanied by an abundance of sensations lying crisscrossed at the confluence of our senses, tickling our emotions and feelings. Following Fontana's artistic gesture, we imaginatively cut deep into time, (ful)filling its past/lost hours with our never satisfied needs.

Fontana's technique serves to show how a simple cut could make paintings three-dimensional, thus helping them overcome the two-dimensionality typically assigned to them. On the other hand, Edson's Vivian gradually became one-dimensional through her painful ordeal. For me, Vivian's tearing of her hospital gown at the end of the play has positive meanings. Only when she tears it apart, only when she is naked again, can she finally break off the cocoon of her hospitalized identity. Seen in the light of Fontana's painting, Vivian *has* to become one-dimensional, because only through this dimension may she be able to reconstruct herself. To better understand this, I propose an exercise: take a pen, place it between your fingers. Draw a line; then another line, then make a cube. The transformation of a line into a cube will thus be more visible. But could we do this exercise backwards? Once the lines have formed a cube, and have become three-dimensional, could they return to being lines?

Having said this makes it more evident why I juxtaposed Fontana's painting to Vivian's experience. When she is totally immersed in the medical equipment, she has lost her individuality. By dropping off her hospital gown, her nakedness attempts to recapture something of her former individuality. In this light, Vivian's gesture becomes paradigmatic for those persons/patients who are on the verge of losing their subjectivity.

Finally, this nakedness is a metaphor for the body-altered/body-in-pain; and yet, somehow, it is still a body-personal, that is to say, not confined in stereotyped, impersonal hospital gowns, nor completely dependent on medical equipment either.

Audre Lorde's *The Cancer Journals*

If Edson introduced a fictitious character diagnosed with cancer, as an example of the enactment of cancer, Audre Lorde's *The Cancer Journals* (1980) documents this real woman's struggle with cancer. In the 1970s, when Lorde was diagnosed with cancer, there were erroneous trends that attributed the illness to the patient/person him/herself. As Diane Price Herndl asserts, "Cancer in the late 1970s was being attributed to a particular—and bad—personality type. Depression, repressed emotions, and succumbing to stress were designated as causes of cancer; therefore the patient was blamed for her illness" (146). Having one breast surgically removed, Lorde finds herself lost, but, at the same time, tries to find persuasive means to communicate her most intimate feelings: "I want to write rage, but all that comes is sadness. […] I am not supposed to exist. I carry death around in my body like a condemnation. But I do live. There must be some way to integrate death into living, neither ignoring it nor giving in to it" (13). Losing a breast to mastectomy provides an opportunity to demystify the myth that a woman is whole only if she is *symmetrical*, narrowly understood as having two breasts. As Lorde recollects,

> In September 1978, I went into the hospital for a breast biopsy for the second time. […] I knew it was malignant. […] The gong in my brain of 'malignant,' 'malignant,' and the icy sensations of that frigid room, cut through the remnants of anesthesia like a fine hose trained on my brain. (27)

Prior to finding out whether or not her tumor was malignant, Lorde sensed fear all over her body. Actually, the adjective malignant seems to have spread throughout her being, obsessively adding a psychical pain to the physical one. According to Scarry, "[t]o have pain is to have certainty; to hear that another person has pain is to have doubt" (7). Yet doesn't Scarry's point of view seem to deny our capacity to empathize with the other?

All her life, and particularly after discovering her cancer, Lorde tried to express her anxieties and stop these "tyrannies of silence" (58). As she writes, reflecting upon her cancer:

> What I most regretted were my silences. […] Death is the final silence. And that might be coming quickly, now, without regard whether I had ever spoken what needed to be said, or had only betrayed myself into small silences, while I planned someday to speak, or waited for someone else to say the words for me. (57)

Her life was changed drastically when she discovered that it could be saved only if her breast was surgically removed. After the surgery, she remembers, "My breast which was no longer there would hurt as if it were squeezed in a vise. […] The euphoria and the numbing effects of the anesthesia were beginning to subside" (38). The phantom limb "[i]s an expression of nostalgia for the unity and wholeness of the body" (73). Consequently, "[t]he phantom limb is not only a memory or an image of something now absent. It is the refusal of an experience to enter into the past" (89).

Although after the mastectomy Lorde encouraged herself by making the rather unusual comparison between her situation and that of the Amazons—those famous mythological female archers whose one breast was cut to allegedly make them more combative/precise in battle—nevertheless she knew she was more than an Amazon. She was a carnal, vibrant, real woman, and not a mythological creature. Furthermore, to her disappointment, she discovered that there were few documents related to other women who had lost a breast to cancer. What happens when women are convinced to wear a prosthesis, and thus fit into a "norm"? In this instance of "normalization," does the hyphened space of doctor-person/patient bear the marks of *scarification* and pressure of those aberrant rules that say a woman is fully a woman only if she has two breasts? As Lorde admits, "To imply to a woman that yes, she can be the 'same' as before surgery, with the skillful application of a little puff of lambswool, and/or silicone gel, is to place an emphasis upon prosthesis which encourages her not to deal with herself as physically and emotionally real, even though altered and traumatized" (89). In addition, "[a]rtificial limbs perform specific tasks, allowing us to manipulate or to walk. Dentures allow us to chew our food. Only false breasts are designed for appearance only" (63). Furthermore, according to Thatcher Carter, "Normalization is the key component in prosthetic breast sales; there is no medical reason to have a prosthetic breast, and the breast is shaped to fit the norms of our society" (665). There are two key words in this succinct, yet powerful passage: "(breast) sales" and "to fit." When women are convinced, after mastectomy that they should return to normality this implies not only an integration into consumer society, but also an artificial reconstruction performed on the site of a female body. But how could such a "breast" be considered healthy, when it is artificial?

This question seems to be addressed by Elise Siegel, too, in her *Portrait # 3* (1992). For me, this work speaks about feelings of entrapment, not only in a gown (i.e., a bra) but also in ideology. Combining wire mesh and modeling paste, Siegel challenges this perspective by remodeling one of the standardized views of a very fetishized item: the female bra. This wire mesh may be decoded as a symbol for us, constantly enmeshed in our spaces of love, hate, ignorance, forgetfulness, and passion.

There is something else about this wire mesh: its filigree-like quality is effectively contrasted by a rigid, black frame. Furthermore, because a bra is a garment worn underneath other layers of clothing, it inevitably provokes another question: What type of image is constructed in a mirror when a woman with a prosthetic breast sees herself? It is problematic for me to assume an answer to this controversial question, especially since it does not result from personal experience. In the context of this chapter, however, I would venture that a prosthetic/artificial breast functions as one of the possible objects with which a body enters into a dialogue. Unlike the objects that retain their own spatiality *vis-à-vis* one's body's spatiality, artificial breasts become identified with the body. Once they are assimilated, their more or less fluent dialogue with the body may suggest that what we see is a body-whole; therefore, the perception of our body is primarily synthetic, and not analytical.

However, the question that intrigues me even more here is not whether or not our bodies can assimilate an artificial breast/organ, but regardless of our health-status, how many times do we reject what is projected onto the surface of the mirror? Those are the very moments when we realize that our bodily contours are as protean as is our uttering of words. Put differently, what is missing when we see ourselves in a mirror is what lies hidden inside ourselves, what cannot be properly put into words, what is deeply buried into the sites of our use of language. Thus, the mirror stage is *staged*, since in front of a mirror we see the (speculative) reflection of our bodily contours. That which is *us* refuses to be present in front of a mirror since we constantly stage, pose, and act in front of it. Of course this play on words (mirror stage*d*) is effective only in English, since in its original French it is image *spéculaire*. However, *spéculaire* comes from the Latin noun *speculum* (i.e., "mirror") which gives English the verb "to speculate." Therefore, to make our bodies matter is to permanently speculate about their meanings, to constantly provoke new readings for our never-fully-developed, always-in-motion-and-becoming bodies. But how could we possibly control our bodies whose shapes are not controllable, but elusive in their nature?

Even more importantly, when bodies are seriously affected by cancer, and when a cancerous tumor is surgically removed (sometimes by excising the neighboring area, i.e., breast, uterus, etc.), and, finally, when persons/patients are pressured to return into a "normality" imposed by socio-cultural and political discourses, we realize that between perceiving the body qua object and the body-abject there exists a very thin line. Allison Kimmich points out that "Abject comes from the Latin *ab*, from, and *iacere*, to throw. Literally, then abject means outcast" (224). In this light, she infers that "The prosthesis would allow the women and the culture at large to deny the possibility of abjection" (228). But are ill women the ones who seize upon this transformation: from object of desire into abject site of repulsion? Or is there a subversive socio-political discourse underneath the surface of this aberrant, illogical transformation? As Lorde writes,

> In the cause of silence, each one of us draws the face of her own fear—fear of contempt, of censure, or some judgment, or recognition, of

challenge, of annihilation. But most of all, I think, we fear the very visibility without which we also cannot truly live. (21)

Therefore, by choosing to alternate two persons in her writings, "I" and "We," Lorde gives expression to the fact that there is no subject in itself but there are subjects affecting each other permanently, zigzagging and thus leaving *visible marks* on the parchment of official discourses. Moreover, if Judith Butler is right when she "[c]onceives of language as a process of reiteration carried forward by the (re)citations of subjects" (Vaserling 27), then when we first see ourselves, and, more poignantly, when we first speak and perform ourselves, there is always a chorus of hidden, sedimented, other voices within ourselves.

Current medicine proposes to reconstruct part of a person's affected body by means of plastic surgery. Ironically, while women who have lost a breast to mastectomy are urged to reconstruct that physical loss from the (fat) tissue of their buttocks or belly, women who have experienced hysterectomy are not expected to do anything surgically to replace the loss of a uterus. Put differently, while there is a false, artificial breast, there is no such equivalent for a woman's uterus. What exactly does this say about our perception of the ill person's body? I think this view reinforces the idea according to which what we see is what is, thus erroneously putting an emphasis on the appearances and not on the body per se, as a whole. Because we cannot see a woman's uterus, we assume it is "there" where it should be. But when a woman is regarded "complete" when she has two breasts, and yet no similar attention is attributed when a woman's uterus has been surgically removed, then we move into a dangerous zone, highly dominated by appearances. The fact that we know how breasts look, but we do not know how a uterus looks, is probably related to our fear of intimacy, lack of anatomical knowledge, and rare appointments to see doctors for CAT scans and/or routine check-ups.

Concluding Remarks

Because medicine is committed to maintaining a sustained level of one's "life-capital," medicine has eventually intervened in how we construct and define our corporeality. During the first decades of the 20th-century, in *Civilization and Its Discontents* (1930), Sigmund Freud anticipated that man would become a prosthetic God, that his essence would be a constant confrontation between conscious and unconscious drives, as well as personal needs and socio-cultural trends. So where is the individuality of one's body? To answer this question, let us mention some examples; when our vision diminishes, we start to wear eye-glasses; when walking becomes too difficult, we use a cane; and when we feel too weak and tired, we stimulate our performance through prescriptive or non-prescriptive supplements. This chapter's section has added another example, no matter how extreme: even paralyzed bodies are kept alive, on "life-support," and hence they continue to function.

Finally, as demonstrated in this chapter through the four examples, it is our moral obligation to reclaim our bodies once they have started to malfunction. This empowering act gives us supreme authority over our own bodies, which, before being clinically diagnosed and categorized as "medical cases," should remain intimate sites, with distinct desires, frustrations and needs.

CHAPTER THREE

BODY-BROKEN/BODY-PROGRESSIVE

In his 1952 classic film *Ikiru* (translated as "To Live"), Akira Kurosawa captures on screen concentric circles of fear and the premonition of death. When Watanabe, the main character, waits for his tests' results in the doctor's waiting room, he cannot escape hearing other persons/patients describing their bodies' symptoms, which unfortunately are similar to his. For example, one says, "Lately, I don't feel myself without a stomach pain." Kurosawa's film may suggest that we often know our bodies' needs and concerns intuitively. Watanabe's doctor, however, prefers to lie to him, saying he has an ulcer. The doctor's attitude is controversial, yet I understand it may be quite difficult for doctors to give someone, whom they barely know, the death sentence. To lie sometimes could serve as an alternative, if we manage to maintain that person's/patient's illusion of his/her former unbroken identity. According to Gadamer remarks that unlike illness, "Health is not something that is revealed through investigation but rather something that manifests itself precisely by virtue of escaping our attention. […] Rather it [health] belongs to that miraculous capacity we have to forget ourselves" (96).

The theme of forgetting (and wakening) is pervasive throughout *Ikiru* and is meant to set the tone for this chapter. Watanabe confronts/sees himself only after he has discovered his encounter with pain. Before this crucial moment, he forgot about himself, having spent all his time entombed at work in papers, papers, and yet more papers. The encounter with his illness makes him vulnerable, and consequently wakes him up.

More than anything, in this film Kurosawa succeeds in proving that an illness, no matter how traumatic it may be, could have positive messages for us. He uses cancer as a basis for the film's plot (if one could say there is a traditional plot here), as well as to show us through it one possible example for the breakage in the habitual body. Similarly, the other persons/patients analyzed in this chapter become aware of the fragility of their embodiments after they have been diagnosed. Therefore, whether healthy or ill, we as subjects are objects in constant mutation, transformation, and possible breakage. In fact, we have been playing with the idea of subjects-qua-objects (and hence broken subjects) ever since we discovered ourselves in front of a mirror, a site of personal *connaissance* and equally *meconnaissance*, just as the concepts of health and illness are difficult to frame in one stable definition.

Memories of the Body-Broken: Why I *Right*

> Some diagnostics are as
> elementary as palpation—
> touching a lymph node lightly
> and sensing its
> swollen from. Other tests require
> that my body be
> illuminated by radioactive
> materials so that my
> inner parts glow on a screen,
> making computerized
> images. Imagism. Imagistics.
> (Rosenblum 11)

We probably first met our double reflected in a mirror, where we initiated a game of gluing together and ungluing the pieces of our elusive "mirror identity." However, whenever we play this game, we rapidly perceive and/or fabricate the changes in our body's choreography, a little bit afraid that if we stayed longer in front of a mirror, we might get hypnotized by our projected double. Another method to analyze our bodies is by looking at photos of ourselves. Nonetheless, this process is less an act of perception than it is more an emotional phenomenon through which we try to recapture the circumstances at the time of the shooting.

It goes without saying that both ways of looking at our bodies are, in fact, examples of seeing their *representations*. Undoubtedly, a small breakage occurs when we face these representations, although this type of breakage resonates in us more on an emotional than on a physical level. Describing Jacques Lacan's "imaginary anatomy," Grosz explains it as "an effect of the internalization of the specular image" [that] "reflects social and familial beliefs about the body more than it does the body's organic nature" (*Space* 86). The clash between the physical map of our bodies and its psychical map may sometimes become irreconcilable. Furthermore, the verbal map of our identity has often been slippery, if not misleading. Put differently, the process of verbalizing our physical and psychical maps has given us the impression we are fabricated, translated, adjusted, and incomplete (if not constantly other) through words.

However, if an illness does not shake our foundations, a preliminary question that needs to be raised is, how frequently do we get acquainted with our bodies? By examining two memoirs (Marjorie Williams' 2005 "Hit by Lightning: A Cancer Memoir" in *A Woman at the Washington Zoo: Writings on Politics, Family, and Fate* and Barbara Rosenblum's 1996 *Cancer in Two Voices*), I argue that, unlike the breakage perceived when we contemplate ourselves in mirrors and/or in photos, the breakage brought about by the onset of their cancer--and its fluctuating side-effects over their bodies--serves as examples of a body-broken. Through experiencing acute pain, their body-broken becomes a body-unknown, a body examined and re-examined, and finally a body that becomes unusable.

Both Rosenblum and Williams write about the breakage in their habitual bodies when cancer was discovered too late by their doctors; consequently, their fight is not only toward winning a battle over cancer, but also a battle with those who were supposed to be able to detect their cancer earlier, when the breakage presumably would have had a smaller impact on their lives. Thus, in their memoirs not only do they write (about) their pain, but they also show how their suffering has the right to be known, for it uncovers uncomfortable truths about the medical practice.

A good venue to start analyzing the problems faced by these women can be through Gadamer's usage of the concept of *en parergo* (or "something which occurs alongside other things") (134). He expands it by suggesting that "We are only able to become aware of ourselves when we are fully occupied with something else that is there for us; only when we are completely involved in something beyond us can we return to ourselves and become aware of ourselves" (134). Is becoming ill, and the implicit rupture that it generates to our lives, a dangerous way to become aware of ourselves? But where would we situate the "return" in this context? There is *the person* (who, up to the diagnosis, has had a habitual existence; who still has friends, job responsibilities, family, and a name); and then, there is *the patient* (who enters hospitals, gets acquainted with doctors and their medical staff, as well as with medical procedures, and becomes respondent, or non respondent, to several treatments). It is the second term (i.e., patient) that seems to gain in importance and capture the first's term space (i.e., person), and thereby generating a discrepancy between a person's/patient's past and current life. Through the experiences within a body-broken, a person/patient drastically redefines his/her identity. As Williams recounts, "I live at least two different lives. In the background, usually, is the knowledge that for all my good fortune so far, I will still die of this disease [cancer, in advanced stage]. This is where I wage the physical fight, which is, to say the least, a deeply unpleasant process" (321). The intimate space of one's person is thus visited, if not invaded, by too impersonal reactions.

Describing how relationships are formed in space, Grosz notes that "Nothing about the 'spatiality' of space can be theorized without using objects as its indices" (*Space* 92). What types of indices does our internal structure possess? The question could be better formulated as, what do these indices exactly tell us as long as they are made manifest on our skin as signs of some advanced internal affliction? It is important to address these questions because they attest the transformations accomplished in time by the art of medicine itself. About these changes, Sontag remarks that

> In premodern medicine, illness is described as it is experienced intuitively, as a relation of outside and inside: an interior sensation or something to be discerned on the body's surface, by sight (or just below, by listening, palpating) […] In the older era of artisanal diagnosis, being examined produced an immediate verdict, immediate as the physician's willingness to speak. Now an examination means tests. And being tested introduces a time lapse […] that can stretch out for weeks. (*AIDS* 35)

In addition to being tested (oftentimes over tested), another difficult task when an individual recovers from a traumatic experience is the act of remembering it; thus, a different type of time lapse emerges. According to Caruth, those who undergo trauma experience "The shock of the mind's relation to the threat of death" which "is thus not the direct experience of the threat, but precisely the *missing* of this experience, the fact that, not being experienced *in time*, it has not yet been fully known" (63). The "I" contained within the word "identity" (or, even better, corporeal identity) finds itself negotiating its new functions and roles. Furthermore, in an era of instantaneity (where we benefit from and are overwhelmed by the usage of digital cameras, cell phones and the internet), it is practically impossible to predict how patiently a person will wait to find out the results of his/her test(s). To rephrase, does waiting intensify one's pain?; does the intellect brutalize the pain?

When in pain, the body—or the affected bodily area for that matter—is thoroughly sensed by us, giving us the impression we are locked inside our body. In this context, Rosenblum provides a pertinent example. Knowing that soon she will reenter the traumatic zone of chemotherapy, she composes a short, matter-of-fact note:

> Dear friends, enclosed is the schedule for my coverage during the second chemotherapy week while Sandy is away. For most of the time during the

> first few days I will be sleeping, so please bring things that will keep you busy. It might be helpful to check in with the persons who will be with me before your shift in case there are things you need to pick up. (18)

This passage is indicative of how Rosenblum meticulously prepares herself before she will be(come) completely unaware of her body during the chemotherapy session's peak. She makes in advance a list of "to do's" because she will be temporarily out of her body. Or, maybe we could argue that she will be completely immersed in her chemotherapy, and therefore not able to function uninterruptedly as before. I also believe she makes this list because her life partner, Sandy, will be gone. During Rosenblum's first chemotherapy session, Sandy noticed certain reactions in her lover's body. In other words, Barbara (the person/patient) and Sandy (the partner) try to create an effective method for dealing with the chemotherapy while being outside the hospital's jurisdiction (At that time, Rosenblum was not hospitalized.). During the first chemotherapy sessions, the two women (along with considerable support from their female friends) tried to find out as much as possible about Rosenblum's body reacting to her cancer treatment. They had hoped her first reactions during chemotherapy would evolve into a pattern, so that they could more effectively minimize its traumatic side-effects. Sadly, no pattern emerged.

Moreover, once her body started to reject all prescribed treatments, Rosenblum was face to face with her collapse. As she writes,

> When you have cancer, you have a new body each day, a body that may or may not have a relationship to the body you had the day before. [...] You don't know from moment to moment whether to call a particular sensation a 'symptom' or a 'side effect' or a 'sign.' It produces extreme anxiety to be unable to distinguish those sensations that are caused by the disease and those that are caused by the treatment. [...] Interpretation of a sensation always depends on having at least two bodily events close enough in time to make meaning of seemingly random events. And most of the time, I live in a world of random bodily events. (166)

This quotation allows a reflection upon the differences between a "body-broken" versus a "body-collapsed"; a body that could still be trusted versus a body that becomes too damaged to attempt any recovery; a body that does not, like all bodies, follow the entropic law, but a body that suddenly succumbs and is consumed by pain. Pasi Falk thinks the human body "[i]s the most obvious and familiar visible 'thing' perceived and yet a blind-spot which tends to disappear in the very act of perception" (1). Both Williams' and Rosenblum's experiences focus on the familiar body turned unrecognizable and, unfortunately for them, unusable. Both are intrigued by their diagnosis, particularly since they both had pursued a healthy lifestyle, had accomplished much in their careers (the former as a journalist, the latter as a sociologist), and had steady relationships. Not only is cancer an illness with an uncertain etiology, but it also throws its sufferers into a crisis of meaning, where the most challenging and frustrating debates happen in the form of fractured corporeal monologues.

On numerous occasions, Williams and Rosenblum remind us that if their doctors did not misread the two women's tests, or, surprisingly, had urged them immediately to have CAT scans, their lives would not have been unjustly cut short. As Rosenblum lets us know, "One fight I'm waging is a malpractice suit against Kaiser Hospital, which failed to diagnose my tumor as cancer over a year ago. They dismissed it, time after time, as benign fibrocystic disease" (20). The process involved in Williams receiving her current diagnosis is even more dramatic. As she recalls, she first sensed a lump during a casual conversation on the phone. When she met with her doctor, he told her, "I would think [...] that what you're feeling is stool, that's moving through your bowel" (310). A second doctor tells her she has fibroids. After an MRI followed by a biopsy, she finds out that "There are at least five large metastases of the cancer in my pelvis and abdomen, and the mother ship [...] surrounding and infiltrating my inferior vena cava [...] Tumors so wide spread automatically 'stage' my cancer at IV (b). There is no V, and there is no (c)" (319).

When Williams' diagnosis was finally given, she was in total shock. She summarizes her appointment with the doctor as follows:

> What it boiled down to was: we have nothing to do for you. You cannot have surgery, because there's so much disease outside the liver. You certainly can't have a transplant—they'll never give a liver to someone who's had extra-hepatic disease; it's against all the rules. You're not a good candidate for any of the newer interventional strategies, and we can't do radiation because we'd destroy too much viable liver tissue. All we can do is chemotherapy, and to be honest, we really don't expect much in the way of results. (326)

Williams' diagnosis, as well as Rosenblum's, put them in the category of "too late," where no recovery was possible. With the cancer spreading more and more inside their bodies, their sense and usage of time shrank considerably. As Williams recounts,

> I have weathered days of wretchedness and pain without a whimper, only to come unglued when some little glitch suddenly turns up to meddle with the way I had planned to use some unit of time: that this half-hour, and the

> contents I had planned to pour into it, are now lost to me forever seems an insupportable unfairness (329).

It is one thing to reflect upon death, yet another to be in the epicenter of the dying process. Living in "Cancerland" (Williams 328), or on an island engulfed more and more by the waters of "it is too late" and "there is not much we could do for you," they try (but do not always succeed) to use their "units of time" as purposefully as possible. Corporeally more dramatically than ever in their lives, they know that soon they will not be able to sense themselves, that they will be deprived of their capacity to use speech coherently, and that ultimately they will not be able to be the subjects of their own bodies.

However, when their attempts at recovery fail to materialize, there still remains the alternative of forgetting. During her first days as a person/patient of cancer, Rosenblum writes in her diary: "My cancer counselor told me that forgetting was the most prevalent symptom among people with cancer. It is, she said, the mind's way to protect you from experiencing the terror of knowing" (31). On her deathbed, she makes another entry in her diary: "I am dying. [...] I have about a month left. [...] Narcotics help. They dull everything a little bit, but the tragedy center of my brain is barely touched by any pill. For that, two stiff martinis and a good cry" (195). Juxtaposed, these verbal images of Rosenblum recording the impact of cancer on her life focus on the possibility of denial (the first quotation) versus her acceptance of cancer (the second quotation). Furthermore, the second entry made me think of the famous Theatre of Dionysus in Athens. In antiquity, plays were performed there, wine was enjoyed, laugher and cry mingled, satyrs were luscious, Maenads were rapturous, and the god of tragedy was thus rightly celebrated. His amphitheatre is now in ruins. Visitors still peregrinate its site, and those with imagination can hear the echoes of much simpler days' celebration of life. Without doubt (and *if* permitted), Rosenblum's story is different. With her body completely destroyed by cancer, literally in ruins, there is yet one single place where illness could not manage to break through. In the center of her brain, the tragedy of her experience plays uninterruptedly. There, less and less frequently, there are also performed recollections of her life as "whole-Barbara," as if her head had become a snapshot of memories.

Does a patient's extreme experience of pain advance more and more or plough deeper and deeper into a person's intimate space? What could be the identity of a patient whose personhood is captured almost exclusively through memories? Finally, does a patient's experiences engulf a person's former identity, who, faced with the breakage in his/her life, cannot resist this change, just like water cannot stay still in a sieve? (It keeps it wet, though, until the water evaporates and disappears into thin air.)

Before considering these controversial questions, it is useful to note Nöel Arnaud's puzzling idea: "*Je suis l'espace où je suis*" (or "I am the space where I am") (Bachelard 137). Arnaud made this remark in reference to what it feels to be standing in a corner. From there, one person would have an angular perspective on all nearby objects; simultaneously, one would feel entrapped, immobile. If we put together these two perceptions, the result may be following into the pattern of transitive thinking. Broadly defined, in the field of logic and mathematics, the transitive phenomenon refers to cases where a new relationship develops between elements that share something in common; for example, if one element is related to a second, and the second is related to a third, then--according to the transitive principle--a relationship is formed between the first and the third elements. In this particular case, when an illness breaks one's physicality, when that breakage is beyond possible repair, and when objects seem to be somehow more alive than those individuals who can barely move because of their pain, then one could feel oneself immobile as an object. At the end of her life, Rosenblum had the same feeling. As she notes, "Although Heidegger is not my favorite philosopher, there are now lessons to be learned from him. A dog dogs. The world worlds. Barbara Barbaras. A world where nouns and adverbs dissolve into essences. Barbara is still Barbaring and that makes me feel alive in yet another way" (128). In this short passage, Barbara has a split vision of her present self: Barbaras (present active) and Barbaring (present participle), both tenses subtly paralleling the presence of pain, which has made her "alive in yet another way," and which, I assume, has helped her reach her "essence." Once she has reached her essence, pain—like the nouns and adverbs mentioned above--began to dissolve. Thus, pain lodged in Barbara's essence, stopped to have a meaning; and it did not matter anymore. As she writes announcing that she lost the battle, "I don't have emotions like you [Sandy] do anymore" (194).

When a body, like Rosenblum's, is so consumed by pain as to be able to feel itself, does this mean pain resides/remains only in the head? Once the human body is close to ceasing its sensations, does this mean it exists only peripherally, that is to say, in the mind? According to Drew Leder, "The brain […] shares with the viscera a depth disappearance. […] Yet this disappearance is nothing but […] the focal disappearance of the surface body. […] As I gaze upon the world, I cannot see my own eyes" (113). We cannot see the brain, but we feel it branching throughout our corporeality. But in those cases where pain is too much for one person/patient to bear, does the mind transform pain into an obsession? Does cognition impair suffering? The brain, the most "absent organ" (Leder 113) from our bodies, seems to be challenged with a very demanding task: it has to make itself *present* to understand pain and its changing intensity and density. The enduring patient's body learns to mediate between *ego sentiens* (i.e., "the sensing I") and *ego scribens* (i.e., "the writing I") which both, through recording fluctuating sensations and inscribing verbal reactions to pain, attempt to discover some soothing explanations to *patior* (i.e., "I suffer").

We frequently ask what pain is, how we construct its meaning, how deeply it damages our tissues, organs and dreams. As suggested by the experiences of Rosenblum and Williams, in the end, following sequence after sequence of bodily deteriorations, pain ceased being felt, or--even better--pain was all that they felt. Hence, to dull their pain, the two women were given strong tranquillizers. Unfortunately, in severe cases, the persons'/patients' bodies may move beyond the initial breakage as caused by the illness, becoming unrecognizable and unusable. Furthermore, the experiences undergone by Rosenblum and Williams reveal that sometimes bodies could break beyond our control, and in that case our commitment to recover them is rarely sufficient. Unfortunately, their bodies did not react consistently to any of the several prescribed treatments. In such cases, it may be useful to "[r]ecover a sense of the importance of minds and cultures in the construction of pain, and we must begin to proliferate the meanings of pain in order that we do not reduce human suffering to the dimension of a mere physical problem for which, if we could only find the right pill, there is always a medical solution" (Morris 290). In other words, to find "the right pill" is symptomatic of a more general and intricate condition, when people, after they have been diagnosed, become "medical cases." The search of finding the most suitable medical treatment to people's specific diagnosis has little to do with offering comforting solutions to their suffering (which, in many cases, involves humiliation and isolation).

Point in fact, there is something else about Rosenblum and Williams that captivated and saddened me at the same time while reading their memoirs. If the two women seem sketchy to readers it may be because they barely wrote anything about their lives before their diagnosis of (advanced) cancer. Using the literary genre of memoir, their writing surprisingly seemed unwilling to unmask a life they had before, although in most cases of physical and emotional breakages, a reconnection with the patient's/person's past is highly recommended. Thus, with no outside narrator to interject him-/herself between what the two women were willing to share with us, their memoirs became a brutal confession of pain, where as a reader one may realize the other, hidden side of the medical practice and discourses, as well as the empty rhetoric of some

health-related campaigns (Almost every product that we buy today has an inscription that imposes its validity through catchy words such as "healthy," "better life," "natural," etc.).[1] Their memoirs constantly suggest the importance of seeing doctors more frequently in an attempt to open a more vibrant dialogue between ourselves and our body. When a person's/patient's body breaks beyond possible recovery, then inevitably medicine's infrastructure is shaken as well. At one point in her memoir, Williams speaks openly about an unpleasant, yet typically encountered truth:

> I try to remember that I am one of the luckiest cancer patients in America, by dint of good medical insurance, great contacts who gain me access to the best of the best among doctors. I'm quite sure that if I were among the forty-three million of my fellow Americans who had no health insurance—let alone a really good insurance—I'd be dead already. (329)

Yet our concern is not only about having or not having (good) medical insurance. We also want immediate answers, just as badly as we want quick solutions to our problems. Part of the explanation of this phenomenon is related to the physicist Isaac Newton who predicted that the Scientific Revolution's "[g]oal is how, not why. That I cannot explain gravity is irrelevant. I can measure it, observe it, make predictions based on it, and this is all the scientist has to do" (Berman 43). In this light, we erroneously see certain institutions as being unbreakable, or immune to break, when in reality they cover nonchalantly their weak points. Simon J. Williams starts his book, *Medicine and the Body* (2003), by reminding us that "Prior to the 'age of reason,' for example, supernatural beliefs about malevolent spirits, ideas about evil and divine intervention, and practices of sorcery and witchcraft were highly influential in Christian Europe" (10). Sadly, there are people who still think that "malevolent spirits" exist among us. The situation gets complicated when we cannot explain the causes of certain illnesses: "The doctrine of specific etiology is likewise problematic—no one-to-one correspondence can be assumed between pathological agent X and disease Y" (Williams 13). When our body is spatialized by an illness, we want doctors to cure it immediately. By so doing, we falsely invest doctors with supernatural powers, demanding them to make the illness disappear, as if doctors knew a mantric spell, or as if they possessed a magical wand and we were living in a wonderland temporarily interrupted by pain. Memoirs such as Rosenblum's and Williams' remind us forcefully that this is not so; that, in some cases, medicine cannot prove its endeavors successful.

As shown in this analysis, human bodies react differently to illnesses, thus maintaining themselves--even through the encounter with their pain--as unmistakably personal as possible. Rosenblum's chemotherapy could not stop her tumor ramifying throughout her whole body's topology. Williams' treatments managed to keep her alive long after her doctors' pessimistic pronouncement (Initially, they told her that she would live 4 to 5 months maximum.). Even more surprisingly, in many cases, her body could engage in its routine as magnificently as it had done before its diagnosis, defying, if only temporarily, the typical damages resulting from cancer. Furthermore, as demonstrated by Sontag in her book, *Illness as Metaphor* (1978), in the 19th century, tuberculosis was considered a scourge for humanity. Still, back in those days tuberculosis was surrounded by a mysterious aura. Some people thought highly of it, considering it to be a symbol of a refined, wondering, Romantic poetic spirit. No such positive connotations are attributed to cancer. Ultimately, this illness has profoundly questioned our confidence that a body, once broken, could be restored through treatments, surgery and rest. In other words, cancer has managed to create a *breakage* in what seems to be situated outside our internal topology, in our socio-cultural and political strata, but which profusely infiltrate into our minds and bodies.

There is yet one more piece that needs to be added to decipher these two women's story according to which "['t]racks of the body,' like other bodily aftermaths: footprints, fingerprints, scent trails, aftertastes [...]. Scars are the body tracks of the surgeon on the patient. [...] Surgery as a discourse overwrites the body's inscription as a cultural text. [...] So bodily inscriptions affect our deciphering of signs of presence in the flesh (Young 86). When people are diagnosed, they gradually relinquish their former selves as they get acquainted with medical procedures and more or less invasive tests. The "I" as fleshed out in autobiographical writings becomes more vulnerable and assertive at the same time. The written (and thus exposed) "I" mediates between the public sphere (as represented by medical institutions and practices) and the private one (as amassed through the experiences recorded beyond one person's epidermal boundaries). The body that was once apparent, *there*, becomes transparent because of acute pain and suffering.

Miasma and/as Uncontrolled Political Discourses

The first section of this chapter introduced the significance of the concept "body-broken" along with its progressively collapse. It is worth noting that people cannot ever completely recover emotionally after they have seen their beloved dying. One thing is certain, though: through perceiving their beloved ones' debilitating pain, they acknowledge the existence of a "body-broken" or "body-other." Furthermore, when a body breaks because of AIDS, our homophobic fears may erupt more violently than we thought they would. Because of the misconception of this illness, we have turned our fears within and, almost exclusively, have paid attention to our erotic lives/history. Our homophobic bigotry reveals how shallow our notions are in regard to how AIDS is transmitted. Moreover, medical treatises bring into discussion iatrogenic complications that could be the result of a treatment's development, thus adding more concerns to a patient's already debilitated body[2]. It seems that AIDS has developed its own iatrogenic enigma, which has not erupted from unforeseen prescribed treatments, but instead it has come through insufficient health-related campaigns.

Anne Hunsaker Hawkins suggestively notes that "[t]he tendency in contemporary medical practice is to focus primarily not on the needs of the individual who is sick but on the nomothetic condition that we call disease" (6). Since a body with AIDS may contaminate a healthy one, a person's illness's spatiality is not self-contained anymore. In an increasingly self-aware society, the medical identity seems to be forceful and resourceful enough to influence other persons' behavioral reactions. The three works analyzed in the second part of this chapter—William Hoffman's play *As Is* (1985), Amy Hoffman's memoir *Hospital Time* (1997) and Hector Babenco's film *Carandiru* (2003)—are a narrative triptych of pain and suffering, patients' isolation and their attendees' limited compassion.

A SwA—an acronym for Society with AIDS—is a terrified community that painfully demonstrates that limited medical knowledge and mass-media sanitized programs that are commissioned by various profit organizations or political parties imprison our minds up to a point when we—"body-broken" or not—search for signs of illnesses all over our bodies, and keep certain individuals, who may be at risk, at a distance. By so doing, we continue to blame homosexuals for an illness whose irrefutable etiology is still not discovered or, even better, scientifically proven.

In Hoffman's play *As Is*, the author captures masterfully a very dark, gloomy atmosphere, as the main action occurs in a hospital. In the preface to the play, the playwright sets the tone and concern of his work: "I had just finished reading the previous day's *New York Times*. [...] I told my friend the article was absurd: a disease [i.e., AIDS] capable of distinguishing between homosexuals and heterosexual men?" (xi) Homosexuality has always been targeted by some as being immoral. As Elizabeth Grosz points out, "In the case of the homosexuals, [...] it is less a matter of who they are than what they do that is considered offensive. [...] Homophobia is an attempt to separate being from doing" (225-26).

Like cancer which is defined by many uncertainties, starting from its etiology to its uncountable treatments, AIDS follows the same pattern. In addition, the latter introduces a breakage situated outside one's body's topology. With AIDS, not only is a patient's body questioned, tested, and investigated, but also his/her partner(s) ought morally to follow the same clinical routine. AIDS may be considered an illness that introduces a breakage in people's intimacy. What was once regarded as safe and personal could become insecure and public. Because of AIDS, Susan Sontag thinks

> Sex no longer withdraws its partners, if only for a moment, from the social. It cannot be considered just a coupling; it is a chain, a chain of transmission, from the past. 'So remember, when a person has sex, they are not just having it with that partner, they are having it with everybody that partner had it with for the past ten years,' runs a [...] pronouncement made in 1987 by the Secretary of Health and Human Services. (72-3)

The play's brilliance resides in suggesting how prone we are to accept lies instead of dealing with uncomfortable truths. For now and then, we make ourselves believe that, as long as we keep doors closed, secrets will not come out; as long as patients are kept inside a hospital, an illness, such as AIDS, can be disregarded; finally, as long as homosexuals are kept "in the closet," they do not exist for us. In other words, Hoffman's play diagnoses several outbreaks, all potent and urgent for discussion/healing.

At least for the moment, Saul—the protagonist of the play—is not hospitalized. He introduces us to his friends, all of whom have AIDS. Teddy "is not in pain" (7) anymore because his body is practically destroyed by the illness; Jimmy, who died recently, had been "in a coma for a month, [...] Harry has K.S. [short for Kaposi sarcoma, which are severe lesions, as one possible complication resulting from AIDS] [...] and Matt has the swollen glands" (8). Finally, Saul's current partner, Rich, is also hospitalized.

The play focuses on two major issues. Written in the early eighties when the epidemic was not yet fully exposed to the public, and when homosexuals were still kept "in the closet," one of the first things Hoffman wants to prove through his play is the ignorance of the authorities. They wrongly misread AIDS as an illness that affects only homosexuals. In so doing, they label them as deviant and dangerous. Even more poignantly, they allow AIDS to proliferate among people who have been told to think their heterosexuality would keep them away from this illness. In the following passage, Hoffman combines eloquently three distinct reactions to AIDS:

RICH. *Doctor, tell me the truth. What are my chances?*
DOCTOR 1. I don't know.
RICH. *Doctor, tell me the truth. What are my chances?*
DOCTOR 2. I don't know.

RICH. *Doctor, tell me the truth. What are my chances?*
[…]
DOCTORS. We don't know.

TV ANNOUNCER. The simple fact is that we know very little about Acquired Immune Deficiency Syndrome. Its victims may live a normal life span, or they may have only a few weeks. Fortunately, so far
this tragic disease has not spread outside its target groups to people like you and me. When will science conquer this dreaded disease? We don't know. We don't know. We don't know. (13)

By paralleling two socio-cultural institutions of power and control, television and medical practice, Hoffman in this passage points out the frustrating uncertainty of this illness' etiology. It is interesting to notice that, while the doctors seem powerless because they admit they do not know how to cure AIDS, the TV announcer rephrases the fear as follows: science does not know how to cure it. Today, AIDS though treatable is still not curable. However, over the years, we have learned important lessons, one of which is that this illness can potentially target anyone. Because of AIDS, the regime of fear spreads from those individuals who have already been searching feverishly for a "contaminated" lover in their intimate, yet broken, chain, to everyone. Thus, it is worth noting that "Cancerphobia taught us the fear of a polluting environment; now, we have the fear of polluting people that AIDS anxiety inevitably communicates" (Sontag 73).

There is sufficient resemblance between Sontag's idea and the remark of Joe, a character from Tony Kushner's 1994 monumental play *Angels in America*: "Freedom is where we bleed into one another" (*Perestroika*, 1.7.37). Kushner's polemic idea has been interpreted in various ways. For this section's purposes, freedom means to have a body able to enjoy life. A "body-broken" rarely has freedom. For issues pertaining to homosexuals and AIDS, Kushner's freedom alludes to the unpleasant reality of still wrongly associating AIDS with homosexuals. In this light, how far are we from the ancient notion of *miasma* (roughly translated as "stench")?

In antiquity, this notion referred to keeping women in labor isolated, for their bodies, being covered in blood and other unpleasant secretions, were considered repulsive and unhealthy. In the history of medicine,

> The miasma theory of disease was prevalent in Europe from ancient times right up until the discovery of microbes. This was the notion that 'bad air'—air that was damp, odorous or polluted—in itself caused diseased. It was believed the sticky miasmal atoms lodged in bodies, wood, fabrics, clothing and merchandise, and could be absorbed through the skin or by inhalation and therefore could pass from person to person or animal to person through contact. (Lupton, *The Imperative* 20)

Is *miasma* theory one of the reasons why we have misjudged homosexuals' style of life? John O'Neill provides one possible answer. He charts the changes undergone from an agricultural to an industrial society (above). In the former, the maternal milk was invested with a special power, for it suggested an indestructible bonding between a mother and her infant. With the introduction of the bottle in the feeding process, that bonding was damaged considerably. Hence, the maternal milk was not considered as fundamental as it had been, and it thus was not regarded as the only means to sustain life in its incipient phase. Moving forward in time, in an industrial society, blood has been invested with life-sustaining attributes (for example, think of transfusions; or of the type of blood an infant inherits). AIDS then contributes significantly to making us cautious of the unpleasant viruses that could be transmitted through blood. As O'Neill argues,

> [i]n an industrial society the social bond may be rendered in terms of the medicalized icon of the gift of blood. [...] For as long as our medical system fails to find a prevention or cure for HIV, we are abandoned to *horror autotoxicus*—the catastrophe of lethal fluids (blood or semen), [...] where the gift of blood has been polluted and now deals death rather than life and love to its trusting recipients. (181-82)

If that is the case, if one's body could experience the horrors of its own toxicity, along with the risk of contaminating others, and if we are a homophobic society, then in his play Hoffman might have alluded to the quarantine effect *vis-à-vis* the patients who are kept in hospital, in most cases until they die. Saul is caught in an epicenter of bad news, friends dying, his own lover in hospital, and—above all—a lot of unhealthy uncertainty. Not only does this illness affect his friends and current partner, eventually destroying them, but it also makes Saul *too* conscious of his embodiment until he forgets what it is to live. He becomes obsessed in his search for clues of AIDS on his body. This is the play's second major achievement. Hoffman captures masterfully Saul in the throes of his own angst:

> SAUL. I have not slept well for weeks. Every morning I examine my body for swellings, marks. I'm terrified of every pimple, every rash. [...] I feel the disease closing in on me. (8)

At the end of the play, Saul flirts with the idea of getting married to Rich, thus literally closing in the disease on him. The wedding rings in this context could be interpreted as the death rings. On the other hand, he may think his proposal could contribute to restoring his confidence, since the institution of marriage has been a symbol of unity and stability for years. Although this institution has major flaws, it is nonetheless a certainty. After he has seen too many of his close friends dying, Saul ardently needs some stable points of reference to help him balance his recently devastated life:

> RICH. My future isn't exactly promising.
> SAUL. I'll take you as is. (72)

The ending of this play, however, is intentionally not explicit. We do not know whether or not Saul will marry Rich, or if Rich would live long enough to get married. As if reading about people dying of AIDS and the introduction of the regime of fear and unhealthy uncertainty were not enough for us to start a polemical debate, at the very end of the play, the hospice worker urges us to continue our discussion by reflecting upon the nature of the endurance of pain, our limits to sympathy, and the worker's delicate job. As he remarks, "They [the patients] get a lot of support at first, but as the illness goes on, the visitors stop coming—and they're left only with me" (76). Along with the hospice workers, nurses "deal with bodies that transgress boundaries that are broken down and violated by illness and medical treatment" (Lupton, *Medicine* 133). The hospital worker and nurses confirm the fact that AIDS deprives patients of their social identity and makes them question their role and usefulness in society.

The second author analyzed in this chapter's section, Amy Hoffman, has never had AIDS; however, because her friend, Michael Riegle, died of AIDS, and because he made her his health care proxy, I read Hoffman's

Hospital Time as a memoir. If the one who writes this work is not the patient, and yet it can still be read as a memoir, it becomes clearer and clearer that between a patient and his/her attendees there occurs an inseparable, intensified identification of criss-crossed experiences. Throughout her book, Hoffman offers uncensored reflections *vis-à-vis* AIDS and the limitations of friendship. The image with which she starts her memoir is that of clocks:

> In Intensive Care a clock hangs on the wall opposite the bed. Big black numerals. One hand that moves in sudden ticks, minute by fucking minute. [...] [w]e visitors are intimidated by all the tubes and wires and monitors that hook up to machines that are hissing, sucking, clicking, chattering. (3)

The clocks are depicted without any embellishment, namely, as they appear on a white wall. There is a symphony of white here: the walls are white; the hospital bed is white; the nurses wear white uniforms. Michael is more and more detached from this world.

Then, as Hoffman asks, how do we react when we see someone beloved by us dying? What do we do when we notice a patient mocked by an illness, trapped among wires and machines that apparently—in an interesting choice of words—chat with each other? One may argue that in intensive care units, the verbal dialogue is replaced by a white noise resulting from a conglomerate of hissing machines. In this sterilized environment, waiting becomes so burdensome that it pains the one who is waiting. Hoffman admits this truth without editing or disguising her anger:

> Time ticks differently next to the sickbed. [...] And there's no comfortable place to sit. The bed is narrow, the patient bristling with needles, electrodes, and other ICU accoutrements that must not be displaced. [...] I sat on the edge of Mike's bed or on a stool or chair next to it. [...] My neck would begin to ache, then my buttocks. My arm would fall asleep. I'd think of obligations elsewhere. I'd get bored. I'd wish I'd brought a magazine to look at. A magazine! Mike was struggling, he has dying, and he needed my total attention. (3-4)

Hoffman faces the impasse of not knowing how to communicate with her dear friend. She and Michael worked together at the *Gay Community*, a newspaper. They were involved in many politically incorrect situations that homosexuality inevitably causes. As an active lesbian, Hoffman shared with Mike, as well as debated, many homosexuality-related topics. As a parenthetical observation, she admits that sometimes she found Michael "crazy," that is to say too passionate and adamant in his opinions. Therefore, when she admits how boring and difficult it is to wait next to the sickbed, she is utterly sincere. In such challenging moments, she does not know if people need courage, or if they should resign their hopes when seeing their friends entombed in hospital machines.

When Michael cannot talk to her, she develops another form of dialogue consisting of flashbacks. The following one relates an episode at a local gym:

> During an exercise we were instructed to try in pairs, the teacher came over to help Mike and me. 'I'm having trouble with this one,' Mike explained to her. 'I've been sick, and I have some neuropathy in my feet. It's hard to balance.' She obviously never heard that word before. [...] She'd never heard nouns [such as]: *neuropathy*, *cytomegalovirus*, *mycobacterial avium introcellulare*, *Hickman catheter*. The AIDS language. (27-28)

But Hoffman never explains the nouns' meanings to the gym instructor, and thus indirectly to us, because these nouns, although now part of the "AIDS language," are confusing for Hoffman, too. This passage reinforces my belief that medical jargon is a closed circuit available only to those trained in its subtleties. This episode also confirms that uttering/hearing some nouns does not mean knowing/learning their meaning, synonyms, and structure. Philosophers of language, such as Wittgenstein, have noticed why we are so tuned into "language-games." We like to play them because, in the middle of a conversation, while someone still speaks, we may anticipate what s/he will be saying contextually. On the other hand, the above cited nouns are too cryptic to allow (many of) us to start a conversation.

Actually, when I was writing down this passage in my notes, and then again upon inserting it here, I was careful to type correctly every single noun. One morpheme changed, and I felt the disaster approaching. As if the change in the nouns' morpheme was actually the big concern here! These nouns represent a series of abstract ideas. It is useful to note that, etymologically, "abstract" comes from Latin, more specifically from the perfect participle form of the verb *abstrahere*, namely, "to draw off." In one of its connotations, abstract means being drawn away from worldly interests. Unfortunately, when a patient's body reaches its end, it becomes more and more abstract; here I use the adjective with the meaning "rigid," in anticipation of the *rigor mortis* characteristic of the dead. Or, as Hoffman writes: "Michael's emaciated body. The heavy, heavy ashes" (86). She must reconcile two different images of her dear friend (before and after his physical collapse), and she knows that will demand courage.

But there is something else implied in this passage. Once Michael is dead, Hoffman will need someone else to minister to her pain.

Retrospectively, she admits she was not much help to a very sick, dying friend. Therefore, she is skeptical that anyone can soothe other people's pain. When Michael was barely present in this world, Hoffman remembers an episode that enraged her so deeply that it made her erupt into shouting:

> 'Hang on. Stay with it,' Loie would whisper to him [Mike], crying and squeezing his hand. She told me, 'He knows we are with him. It helps him get through the pain.' Sorry, but I can't believe it. Get through the pain—to where? More pain? Mike wasn't going anyplace. He was just dying, on his deathbed. He had no relationship left, except with pain. (123)

She thinks that some wounds remain open; with the passing of time, they may subside in intensity. Sadly, she admits that she could not find any other way to deal with Michael's suffering that has now become, to a certain extent, her own.

Continuing this very open discussion about AIDS and homophobia, Babenco's film analyzes the extreme representation of illnesses. "Carandiru," where the action of the film occurs, is a prison in São Paolo, Brazil. The characters face a double imprisonment: one literal, in prison; another metaphorical, in their bodies with AIDS, tuberculosis, or scabies. Babenco suffocates us visually with sick inmates. There are so many, they are not properly identified throughout the film. As one of them says to Dr. Varella, who is conducting social work for AIDS prevention at "Carandiru," "I got AIDS, is it still worth taking the test?" Another inmate says, "You come in here sick, they treat you with respect."

Throughout his film, Babenco does not offer answers related to promiscuity and the trafficking of illnesses in prisons. By its end, we do not have the feeling we know these inmates any better. In fact, there is no sense of identity in this film; there are no main characters, no plot, or action. Everything happens, or lulls, in front of the camera, as if there were never an initial script. Babenco shoots the ordinary, regimented routine that exists in prisons, giving the impression he moves his camera from one scene to another dispassionately. According to Sean Cubitt, the "[c]inematic present […] can be given a number: zero. […] The concept of nonidentity reveals zero's quality of internal difference. Zero is a relation rather than a (no)thing because it is always a relation of nonidentity with itself. Zero acts, rather than is because of this instability. And it acts in relation to the cardinal numbers (1, 2, 3)" (33). If zero defines its identity through the presence of the other cardinals, then zero possesses this remarkable quality of being in perpetually re-constructing identity. Apparently, Babenco's film is populated with zeros, meaning the inmates, and one cardinal, Dr. Varega. But the slim language of numbers is elusive.

On closer analysis, it is Dr. Varega's identity as an outsider that is questioned. He becomes zero in this situation, although here I use zero with a different connotation. Once he completes his social work at Carandiru, he will stop being in contact with these inmates. His relationship, or maybe I should say interaction, with the inmates is episodic, and hence rather fruitless.

For reasons not clearly stated in the film, in the end these inmates are killed in bloodshed. The extremely graphic image of the dead could be interpreted as another way to illustrate the cessation of suffering. Both their illnesses and wrong actions are cleansed in one violent act, so that we see more clearly the metaphor of quarantine upon which this film is created; people with AIDS could contaminate healthy people, and—if possible—they should be kept isolated. Have illnesses, and suffering in general, somehow exhausted our patience?

To answer this question, it is worth mentioning that the process of waiting has different meanings for the patients and their attendees, particularly when the former are prescribed tranquilizers. As Larry Dossey conjures, "[w]hen we experience a technique that diminishes pain through expanding our time sense, we are not merely exercising self-deception. We are not fooling ourselves into thinking the pain is not there" (47). If the awareness of time for the sedated patients is altered (that is, they do not have a clear sense of circadian time anymore), time for their attendees seems to have stopped. Although they are not physically affected by their beloved's illnesses, nonetheless their lives are at a halt because of pain. To better understand this, let me return to Sontag's idea according to which, "Compassion is not a stable emotion." Sontag argues further that, "It [compassion] needs to be translated into action, or it withers" (101). As someone who has been involved in observing, and, whenever possible, helping two of my relatives deal with their pain and eventual death, it is important to say that compassion is never enough. Compassion is like a drug with too quick an effect. When one cannot do anything *more*, one feels helpless. One feels s/he is companionless to the one who is in pain, but never compassionless. Hoffman, whom we have noticed giving such honestly brutal accounts of her being bored and tired of her visits to the hospital, writes in another entry of her memoir: "By that last week, I knew the way to the hospital by heart, I knew millions of ways. I thought about Mike all the time, and I couldn't sleep or enjoy sex, food, work, companionship. Only in the hospital, looking at him, would my thoughts of him leave me" (67).

Furthermore, compassion may not mean much to patients on their threshold of death, who become more and more detached from such an "unstable emotion." In hospitals, or at home, when an illness is too advanced, baffled we could say: "Illness itself is a strangeness" (Zaner 36). The suffix "-ness" typically describes an object's essence. But could we attach it to people who are close to death? For Hoffman, at the end of her friend's ordeal, Michael became "Michealness" (142). Moreover, when bodies collapse, when treatments are not compatible with the patients' bodies, and when there is no satisfactory meaning to their ordeal, time itself seems to freeze in its "time-ness." So what does one say or do when time reaches its time-ness, when the beloved is close to death?

Illnesses, like musical, literal and political trends are reflective of the time in which we live. Like cancer, AIDS is not one single illness, but a complex of many intricate symptoms that baffles the medical community and raises illogical levels of fear in us. The ancient Greeks and Romans believed in the humoral theory in which "the environment, in combination with individuals' constitutions, were influential in affecting people's state of health. The humoral theory of disease incorporated an understanding of the healthy body as maintaining a balance of the four humors, blood, phlegm, black bile and yellow bile, four elements, earth, air, fire and water, and four qualities, hot, cold, wet and dry" (Lupton, *The Imperative* 19). This type of embodiment was closer to nature and more prone to accept illnesses as curses that needed their cure or redemption in a divine intervention.

On the other hand, an unbalanced body because of AIDS becomes "miasmatic," that is, a hideous, repulsive and isolated body. This is a consequence of incorporating without filtering all those uncontrolled political and medical discourses, according to which "The image of the 'positive' body or the body with AIDS is strictly controlled. Nowhere is an image of the 'ugly' or diseased body evoked directly, for any such evocation would refer back to the initial sense of AIDS as a 'gay' disease [...] *Mens non sana in corpore insano* cannot be the motto" (Gilman 162). Bodies with AIDS bleed differently than other bodies. Their blood is poisonous and vengeful. If touched, it may contaminate another. A contemporary disturbingly distorted myth of hygiene is born; a body with AIDS echoes the myth of Medusa. That is to say, a body with AIDS does not have the legendary force to decapitate those who look straight into the monster's eyes; however, no one can deny that a body with AIDS unwillingly transforms a simple, spontaneous touch into an irrational fear.

Whereas "smell and touch [used to] evoke for us the world before language [since] they [were] keys to repressed memories of the wholeness of the world not primarily seen but felt and tasted and smelt" (Gilman 178), with the official acceptance of AIDS as a contemporary medical conundrum, the act of touch has been removed from our intimate sphere and may, in some cases, become litigious. Furthermore, although *miasma* theory was abandoned when the theory of microbes was discovered during the closing years of the 19th-century, it has nonetheless managed to infiltrate itself into our current hygiene-related campaigns. For Lupton, "In the case of the 'new' public health, individuals are largely governed through inciting them to exercise personal autonomy and political awareness. […] Thus, the 'new' public health […] demands even wider hygienic strategy [in which] every individual [has become a master in] the techniques of self-surveillance" (*The Imperative*, 76). While it is politically incorrect to conceive of AIDS as a "gay" disease, unfortunately few changes have been made towards not keeping these patients isolated or perceiving them as pariah. A new political barrier has been erected that divides us into the social category of those who are considered "safe" (and with "rights") and those who are viewed as contagious (and deprived of complete control over their bodies or public identity).

Conclusion

Addressing how the dynamics of language could be changed if it was conceived from a verb's, and not from a subject to object, perspective, David Bohr "[h]as proposed a new model of language called the 'rheomode,' emphasisizing the Greek word that means 'to flow.' He suggests that a primary role be given the verb instead of the noun, thus reducing the emphasis on subject and object" (Dossey 204). Needless to say, the verb was employed in this new dynamic of language because of its capacity to suggest and express action. But verbs get locked into their "-ness," too, when there is not much left to do and/or offer to a patient and his attendees. When the verb freezes too, ironically then we hear again the burden of time's ticking. As Dossey argues, "We wear a watch with no conscious regard for the name we give it. […] Using it, we *watch*. We watch time, we are fixated on it. […] constantly watching, always watching, it is we who are in the service of time" (29).

This chapter offered a reversal on the perspective of pain, according to which it extends from the one in literal, corporeal pain to his attendees. After we have taken care of ill people, we wear a different type of inscription tattooed on our bodies: "Beware!" But I need to ask: Beware of what: of pain? of physical breakage? of emotional collapse? of death? Could we watch our bodies as if they were outside of us, ready to pour down on us their illnesses, deviations, and misfortunes?

In reaching the conclusion for this chapter, we could interpret a collapsed, "body-broken" being similar to a convex mirror where nothing stands still and where there is no stable center of gravity anymore. A body in acute pain has various nociceptive foci. On such a mirror, a person/patient perceives her/his new corporeal identity as being created at the intersection of two axes: the "horizontal body" (which is confined in bed day, after day, after day) and the "vertical," vigilant mind (which retains a hope of the body to being restored to its previous condition). Finally, we may interpret these two axes as a cut, a line, or a breakage inserted in one's identity because of intensifying pain.

Notes

[1] I envision that, in years to come, these labels will function as cultural "hieroglyphs" of our limited information and/or knowledge in regard to human anatomy.

[2] As defined by the dictionary, iatrogenic techniques are induced in a person/patient through therapy, resulting in an infection or any other unforeseen complication of the initially prescribed treatment.

CHAPTER FOUR

AT THE EDGE: THE OTHER'S LIMINALITY

Comedy and pain both share the body as their
common ground. In fact, bodies—from a
comic point of view—are almost inherently funny.
(Morris 81)

Dying's not just about you!
(Lavery 71)

If so far in this book, we have focused almost exclusively on the person's/patient's diagnosis, coping strategies and reactions to it, and if we have witnessed several bodily breakages, this chapter proposes to take a look at the way family and friends react to shocking news. Along with the physical liminal space (as it belongs to a person/patient), there is also the attendee's highly delicate emotional liminal space. Put differently, in ancient Greek drama, a chorus was an emblematic figure of old and wise men who had the power to predict the consequences of some wrong actions, or the lessons to be drawn out of greedy and/or inappropriate behavior, along with their impact in a community. In this chapter I reflect upon the notion of the chorus in regard to persons'/patients' illnesses, or in some cases, already non-communicative human beings with practically inert bodies. Therefore, how profound is our implication in such cases? On the other hand, how devastating is it for us to see our beloved ones dying? What is the difference between *esse* ("to be") and *percipi* ("to perceive")?

In her book, *The Body in Pain: The Making and Unmaking of the World* (1985), Elaine Scarry asserts: "Because the person in pain is ordinarily so bereft of the resources of speech, it is not surprising that the language for pain should sometimes be brought into being by those who are not themselves in pain but who speak *on behalf* of those who are" (4). Needless to say, there is a clear distinction between incipient pain and its last stages, where one's language as well as one's body's stamina are devoid of power and significance, rendered almost, if not completely, unspeakable. However, speaking on the other's behalf is never easy; explaining the other's pain is misleading. One translates the other's feelings, and, to a certain extent, one intervenes verbally in that experience.

This process creates a painfully sensitive interface between a person/patient and his/her relatives. If permitted, the experience of pain could be compared with John Cage's musical experiment where the player, his instrument, and the audience, *all together*, are creating that unique, irreplaceable *4'33''* (1952). I alluded to Cage's musical experiment because of its boldness in transforming the audience into *ad-hoc* players. Similarly, although with magnified sensations, our beloved's pain makes us *ad-hoc* sufferers? The question is how tactful and prepared we are in mediating their pain. Sitting next to a patient, his attendee faces the burdensome ticking of clocks, the ache of waiting, and the dagger-piercing questions of one's meaning. In other words, it is not only the pain of the other that intrigues and baffles us; it is also experiencing, internalizing and then performing our reactions to that pain.

Vogel's *Baltimore Waltz*

When dramatists approach the motif of the body in pain, it is extremely challenging for them to offer a believable production that equally respects the character's pain while meaningfully connecting it to the audience. According to David Haradine, "In performance, the body, or traces of the body, or echoes of the body in its absence [...] are the foundations upon which the very notion of performance is predicated" (69). Contemporary dramas written on the subject of pain face an even greater problem. The cathartic moment, as a former *sine qua non* element for tragedies' dénouement and climax, seems to have been removed altogether. In so doing, neither the character who is in pain, nor his audience, could achieve that sense of temporary release from suffering.

In Paula Vogel's *Baltimore Waltz* (1996) and Bryony Lavery's *Last Easter* (2004), the dramatist proposes a radical version of the theme of waiting where people are exhausted by their (passive) waiting to find a cure for AIDS. More specifically, after the shock of the diagnosis, and especially after many treatments have not given results, persons/patients have the option to accept their fate by smiling at its cruelty, or, as David Morris contends, "As medicine will attest, the possession of a body absolutely guarantees the comic prerequisite that sooner or later something will go wrong, often painfully wrong" (Morris, 81).

With a remarkable sincerity, in the preface of her play, Vogel confesses about a trip to Europe that she was supposed to take along her sick brother.

She postponed it, and, unfortunately, her brother died soon afterwards. *Baltimore Waltz* is a memory play that allows her to commemorate her brother's death and, at the same time, travel imaginatively to Europe. Thus, whenever Carl (her brother) appears on stage, he is sketched out from memory and, hence, slippery. As motto for her play, she quotes a character in David Savran's *Breaking the Rules*: "I always saw myself as a surrogate who, in the absence of anyone else, would stand in for him" (6). This explains why Vogel does not have a partner to waltz with, except in her vivid imagination and recollection.

Describing the main technique used by Vogel, Savran believes she follows Bertold Brecht's "the alienation effect," derived from the Russian formalists, who claimed that "[o]ver time our perceptions become increasingly habitual and automatic: we no longer see what is around us. The purpose of art is to restore visibility, to *defamiliarize*, the commonplace so that we notice it again" (xi). A waltz is danced with a partner, who, in Vogel's case, misses, and will forever be absent. But she dances it anyway, tracing backward the agony of her brother's illness, of which, at that time, she regretfully knew so little.

Vogel's play might profitably be situated between two other better known dramas on the subject of AIDS, Larry Kramer's *The Normal Heart* (1985) and Kushner's *Angels in America: A Gay Fanstasia on National Themes* (1994). In the former, the patient with AIDS is unofficially wedded to his partner at the end of the play, while Kushner dreams that the next necessary "Perestroika" will be that of accepting homosexuals and speaking openly about AIDS. In her play there is no compromise reached through a counterfeited happy ending. Vogel constructs her play allegorically; the patients have ATD, short for "The Acquired Toilet Disease." Anna, the main character, is Carl's sister, and he has ATD. If she found out about this illness by mistake (namely, only after her brother was diagnosed), then the implication is that few people know about ATD. She thinks they have been robbed of vital information and urges this situation to be remedied immediately:

> ANNA. Why hasn't anybody heard of this disease?
> DOCTOR. Well, first of all, the Center for Disease Control doesn't wish to inspire an all-out panic in communities. Secondly, we [doctors] think education on this topic is the responsibility of the NEA, not the government. (11)

Since nobody wants to assume responsibility for making this affliction public, for the moment it is recommended that Anna should follow the Public Health Official's dispassionate advice:

> *Here at the Department of Health and Human Services, we are announcing Operation Squat. There is no known cure for ATD right now, and we are acknowledging the urgency of this dread disease by recognizing it as our eighty-two national health priority. […] The best policy, until a cure can be found, is of education and prevention. Use the facilities in your own home before departing for school. […] If absolutely necessary to relieve yourself at work, please remember the Department of Health Services' ATD slogan: Don't sit, do squat!" (19, emphasis added).*

Although ATD is not considered a medical priority, ironically, it has already been labeled. Let us consider the other significance of the word "squat," which is part of a colloquial expression, "to know squat about something." "Squat" may be another derogative word for indifference, disrespect, and intolerance. Thus, considering its nickname, "Operation Squat," we may now know why this is not a priority for health officials.

Looking backward, when cancer permeated our society, we were led to believe it was an affliction generated primarily by pollution, improper lifestyle and depression. Women seemed to be the main target of cancer. A similar fallacious reasoning is employed for ATD/AIDS. Apparently, according to the doctor in Vogel's play, there is a special category of people at risk here, namely those who are single, may come in contact with single people, and are drug addicts.

Interestingly, while developing her plot, Vogel does not place an emphasis on the chain of lovers, as the subject of AIDS has typically been rendered fictionally, but on hygiene, since this pertains to all of us. It is not about an illness per se, as it is about a community and its (lack of proper) education. Societies that are deprived of information are primitive, barbarian, and vulgar. Here Vogel proposes a critique of advanced societies that are not quite properly informed and educated. There is no complete freedom when people are deprived of accurate information. Hygiene is one important venue through which health campaigns become effective for the lay community. However, AIDS is *not* a hygiene-exclusive affliction, since that reasoning would oversimplify its etiology.

When analyzed closely, there exists a good deal of irrationality in these hygiene-related theories. This is why Vogel's "Don't sit, do squat" is ridiculous. If one came upon only this piece of order/advice, one would not possibly take it seriously. Instead, one would imagine it to be part of a more elaborate new book of manners. Furthermore, Carl, who has ATD, is actually absent from the play. He is more like a ghost than a fully developed character. His voice is not directly heard. By the time this play begins, his sister has already appropriated his role, as she attempts to understand Carl's smothered agony.

This explains why health advice is directly targeted at those who are not yet ill and whose arrogance over their invulnerability should be addressed. Vogel suggests sarcastically that, as long as we "do squat," we are safe; as long as we follow orders, we are protected. Needless to say, this type of education and prevention is damaging because it does not offer a comprehensive point of view. In other words, the surveillance regime has moved to another level, in part, because of misunderstandings about AIDS, where people tend to be concerned almost exclusively with how this illness is transmitted.

Point in fact, Vogel develops her play's argument around the trope of sitting/being seated, which can be decoded as a symbol of the ignorance of self-comfort. As a patient in her play admits, "You've got to watch were you sit these days" (13). Janet Stein tackles the same theme, only visually. About her work, *Queen B. Easy Chair Dress* (1989), she writes: "I'm interested in the metaphoric possibilities of a well dressed chair, and its beckoning promise of comfort. However, I created a hybrid and elevated it. It's been given an interior and I've crossed the language of shelter with that of the human façade" (qtd. in Goldin 28). Stein transgresses the meaning of AIDS by considering it a social discomfort generated by immense "doses" of mainstream arrogance and self-protection; by contrast, and without discrimination, Stein believes sitting on a chair may relax everybody's tired muscles. Both Vogel and Stein reflect on how illnesses reveal some of our deepest fears and phobias. In return, yet independently of each other's project, they decide to treat this obsessive fear contrapuntally, that is, deplorable and ridiculous.

If extreme dark comedy is the new genre that defines our lives, then an entire repertoire of tragic heroes and existential motifs would need to be modified. For example, Sisyphus has never been depicted as laughing (and, thus, generating a cathartic outlet) when the unfriendly stone kept on rolling back down. Therefore, we always assumed he was crying and cursing, and immediately associated his stone with our own recurrent worries and anxieties. Based on Vogel's indeterminate reading/approach to a serious, incurable illness and our reactions to it, the lesson learned is that we should approach our misfortunes from a new angle. We either add the element of laughter to Sisyphus (and by extension to our suffering), or remove him altogether from our imagination. An emblematic figure that keeps reminding us about our misfortunes, Sisyphus, unfortunately, does not provide comfort for our troubled, tired minds. Vogel envisions a radical solution to this problem, since her play implies that we should be brave enough to find the laughable meanness and illogical or inexplicable etiology of some illnesses and, consequently wear "[t]he comic mask, which is ugly and distorted but causes no pain" (Morris 85).

A distorted, comical mask-effect involuntarily becomes the result of those episodes when doctors are not sure of a patient's diagnosis and/or of a proper treatment. This is how evasively the doctor explains his findings:

> There are exudative and proliferative inflammatory alterations of the endocardium, consisting of neurotic debris, fibrinoid material and disintegrating fibroblastic cells. [...] Also known as Löffler's Syndrome, i.e., eosinophilia, resulting in fibroblastic thickening, persistent tachycardia, hepatomegaly, splenomegaly, serious effusions into the pleural cavity with edema. It may be Brugia malayi or Wüchereria bancofti—also known as Weingarten's syndrome. (Vogel 9)

There are so many technical words here (both real and invented), but, in spite of them, the doctor is uncertain when giving a diagnosis.

If we go back to the beginning of the play, Anna clutches "the Berlitz Pocket Guide to Europe" (6). Before she arrives to Europe, she has started rehearsing some standard expressions that may come in hand during her travel; for example, "Help me please." (*Recites from memory*) Dutch: "Kunt U mij helpen, al Stublift?" [...] "Where are the toilets?" "Wo sind die Toiletten?" (7) Any foreign language appears unnatural to those who cannot understand it. Although there are familiar sounds in many languages ("m," "t," "s," etc.), each has its own distinct lexicon.

When performing a foreign language in front of a native audience, any slight mispronunciation of a word may become entertaining. We may want to utter one word, but the outcome may be altered. Similarly, we do not understand what the doctor tries to transmit us. In this case, the ill body has become a foreign site and the doctor its confused and confusing interpreter. Through this contrapuntal note, Vogel captures the downsized effect of a serious affliction when it is explained—or should I say distorted—through such bombastic terms.

To make her argument more potent, Vogel adds yet another example. She presents a similarly comical point of view that, this time, belongs to a practitioner of alternative medicine, Dr. Todesrocheln. He lives in Vienna (the birthplace of waltz) and is a "practitioner of uriposia, he drinks the urine" (15).

> DR. TODESROCHELN. We must have many more such specimens from you [Carl]—for the urinocryoscopy, the urinometer, the urinoglucosmeter, the uroacidimeter, uroazotometer, und mein new acquirement in der laboratorium—ein urophosphometer" (52).

It appears that even to be offered healing through drinking urine is not that effortless as we may have first thought. Just like traditional treatments, drinking urine implies having been (over)tested, and, sometimes, more tests bring about contradictory readings and more ambiguity. In reality, this type of medical approach does not always seem to be the rule. Oftentimes, people visit doctors who, perhaps superficially, prescribe a textbook course of treatment, and, if that fails, another may be recommended.

On the other hand, from what we have noticed so far, either from the doctors' evasive discourses or from Anna's exhausting waiting, there is a constant deferral of meaning. No one knows what ATD is; hence, no one attempts to prescribe a definitive treatment. Vogel desperately searches for validation, so that she could finally have closure in her mentally exhausting post mortem relationship with her brother.

Furthermore, this play's plot invites us to deconstruct an ontological paradox. As humans we cannot escape suffering. But how much suffering could we actually endure? The best way to rephrase this question is: how much water could a fish stand? We may go a step further in this parable by arguing that a fish stops needing water once it is out of the tank. By comparison, our endurance of pain ceases only when we are dead.

One of the reasons why we cannot easily move forward in life results from our static waiting. We wait for something else or better to happen. When Anna realizes that there is not anything left to do for her brother, she admits: "The problem with being an adult is that you never forget why you're waiting. When I was a child, I could wait blissfully unaware for hours" (52). Although she does not say it explicitly, now she is waiting for her brother to die. Even Anna learns to mask her suffering and anxiety in an attempt to ease her brother's pain.

On a larger scale, at the end of the play, we could visualize people doing squat, immobilized in that irrational, but allegedly risk-free, position. In other words, one *has* to "do squat" to feel the insanity of that slogan, and these are the deplorable and laughable dimensions expressed in Vogel's play.

Moreover, people seem irrationally eager to attempt any treatment, when traditional ones have proved ineffective. As Harry Lime, a minor character, asserts: "When they're desperate, people will eat peach pits or aloe or egg protein—they'll even drink their own piss. Gives them hope" (50). We notice here another contrapuntal remark, where hope is constructed as a combination of desperate needs and means. This is also a critique addressed to those health campaigns, according to which we are/become what we eat. However, when choosing to eat healthy, we may develop a satisfactory, yet empty, placebo-like feeling. In other words, we have removed the epicurean component from eating, maintaining us instead in a drastic regime of healthy snacks, healthy drinks, and healthy habits. The obvious question is to which extent we have developed a healthy conscience of what we are doing to our minds and bodies.

Very long time ago, the alchemists discovered the *elixir vitae* that, presumably, had the miraculous capacity to renew one's body because it was an agent of bodily renewal. Comparatively, Dr. Todesrocheln elaborates his theory: "Let us look at the body as an alchemist, taking in straw and mud und schweinefleisch and processing it into liquid gold which purifies the body. You might say that the sickness of the body can only be cured by the health of the body" (53). In his vision, the body is simultaneously the agent and agency of its renewal, transforming itself into "liquid gold." This echoes some current health campaigns that promote the urgency of cleansing our bodies off their excesses, toxins, etc.

To illustrate this idea better, let us return to Dr. Todesrocheln's theory, and analyze it against the initial slogan proposed by the play. Doctor Todesrocheln is a homeopath for whom the body is a holistic site, and consequently it is capable to restore its balance through whatever originated its pain/imbalance. The solution of

non-traditional medicine is to drink piss; the choice of prophylaxes is "do squat." Neither is a valid answer and we notice, once again, Vogel's powerful contrapuntal technique where she creates an imaginary dialogue between divergent types of medical approaches. Her hope is that just as a waltz is gracefully danced with a partner, traditional and alternative medicine should collaborate more effectively, so that we may benefit from their improved findings.

Lavery's *Last Easter*

Vogel's play ended with an uncertain, yet controversial, view in regard to the role of medicine, traditional or alternative, as well as the efficiency of health campaigns. On the other hand, Lavery's *Last Easter* (2004) starts ambiguously. The playwright describes her persons/characters as appearing "[t]o be in a rehearsal for something [...] They overhear, watch some scenes they are not in. When not in scene, they get on with their lives...working, eating, sleeping [...] The actors create all the places [...], as if rehearsing in a bare room" (1). June, the main character, has cancer, and her chances for recovery are nonexistent; as she says, "[n]othing is working. The Bristol Diet. All Western medicine. All Alternative medicine. Positive thinking" (17). Nonetheless, her friends will take her to

Lourdes; as hearsay has it, sick people miraculously become well again there. Although June knows she will not recover from her cancer, she agrees to go on pilgrimage because she wants to meet her God in pain, or make God suffer just as much as she has been suffering.

For this dramatist, the theme of waiting is a ridiculously overused cliché, and it uncovers the unnecessary anxieties added by our waiting to find a cure for cancer. Lavery suggests we stop waiting; instead, we must use and accept sarcasm as the best-at-the-moment alternative. Undoubtedly, this sarcasm is a form of agony. Yet maybe it, in the form of vaudeville routine, is the best, although underused, analgesic for our pains:

> GASH [to LEAH]. This man. Goes to the doctor's…he says… 'Doctor, doctor…I keep thinking I'm a pair of curtains…' The Doctor says…
> BOTH. 'Oh… pull yourself together…'
> GASH. This man goes to the doctor's… he says… 'Doctor, doctor… I keep thinking I'm a pack of cards…' The Doctor says…
> BOTH. Sit down… I'll deal with you in a minute… (5)

Furthermore, along with mocking the theme of waiting, as well as introducing medical genre jokes, Lavery proposes to deconstruct the theme of illusion. Since June did not find a cure for her cancer at that miraculous site, she does not comprehend either why Jesus was abandoned by God, *unless* "He must have known who was going to betray… [who] was going to stand by […] so he could run the whole thing to his entire satisfaction… and make it a really memorable and beautiful Easter" (44). What is implied in June's wretched dissatisfaction is a variation upon the act of betrayal. Intuitively, Jesus knew who would betray him, and, more importantly, why. We do not know *what* will betray us (or *when* our bodies will undergo a major change/breakdown). The first act of the play ends abruptly, and until we read the second act, we do not know if we have been transported into the Garden of Gethsemane or if we are still in the company of June's friends.

In a contrapuntal technique, the second act is about pain and dying that have been accepted, or as June says, "[o]nce you do devote yourself to the business of dying, that's a fascinating project too!" (70) The first act becomes the basis for the second, where we see not only June's acceptance of her death, but also her being developed into a character by her actors-friends. Because June's body and mind have reached insurmountable levels of physical and emotional endurance of pain, she realizes that there is only one way out: not through slow death, helped by tranquilizers, but through suicide. To accomplish this, she needs help. In her vision, she has to wake God up and make Him participate in her drama. To do this,

Lavery makes Gash, June's friend, an accomplice in her death (The meaning of this character's name acquires more depth once we read it as in "Oh, my Gosh"). Then, once Gash becomes involved in this criminal act, he may bring some hope and comfort to other people in pain. The initial lines of the dialogue between June and Gash include this interchange:

> JUNE. [t]o do what I tell you… whatever happens…
> GASH. I thought it was to do with the choice of music at your funeral!
> JUNE. The oath was unspecific, but binding… (55).

But Gash is scared; how could he assist June in killing herself, even if she does not have any chance for recovery? This act may be viewed as an extreme form of gratitude to show a friend. Tormented by doubts, he wants to forget what just happened and tries to engage in a casual conversation with Leah:

> GASH. Let's do what we normally do.
> LEAH. Okay. What?
> GASH. Let's say, 'Yes, we'll do it.' And hope something happens so we don't have to do it.
> LEAH. What could happen so we won't have to do it?
> GASH LOOKS AT HER [June].
> Oh. Oh. [meaning waiting for June's natural death] (64-65)

For someone who is a believer in God, or for those who have lost someone beloved, Lavery's solution may appear irreverent. They may feel entitled to ask, if pain is too unbearable, then is the acceleration of death the solution? In his book, *The Loneliness of Dying* (1985) Norbert Elias raises two similar questions: "What does one do if dying people would rather die at home than in hospitals, and one knows that they will die more quickly at home?" (91) Elias poses these questions to let doctors know that the patients' relatives and attendees would like to assist their loved ones more. Explaining the title of his book, Elias reaches a sad conclusion. He believes that, due to the restrictions imposed by hospitals' policies and their ethical etiquette, the patients are bereft in their experience of pain and denied much needed comfort from their families and friends.

On the other hand, Lavery's play seems to suggest that in some cases it is proper to help pain cease permanently. Her play uncovers its meaning when it is read next to its title, *Last Easter*. Here the adjective "last" announces an end to the ritual of the death of Jesus followed by his resurrection. There is no "after," or sense of soothing atmosphere in Lavery's play unless we are ready to accept uncomfortable truths and/or unorthodox methods (such as assisted suicide) in dealing with extreme pain. In order not to consider Lavery's theatrical gesture blasphemous, one must know what precipitated June's decision to kill herself:

> JUNE. Been reading an article… Quite interesting what's happening in the Netherlands at the moment…
> GASH. *Netherlands. Ding! This is going to be something progressive and difficult, isn't it?*
> JUNE. In this doctor's office…in Amsterdam… there's a teeny-tiny bronze statue of a young girl. […] It's of … this girl with bone cancer. […] The disease progresses, she thinks…'if my little legs go, I will go' [i.e., commit suicide]. Her little legs go, she thinks, 'hey, I can still play guitar…talk to my friends... If I become incontinent, then I will go…' Now she's incontinent but she thinks 'I'll use nappies. That's okay.'
> GASH. *The Human Will to Live. A magnificent Story Arc! There's a Big Play in It!*
> JUNE. Then… the doctor does a house call… the little girl she says 'something's going in my brain. That's enough. I think I will die tonight.' The doctor says I understand, there's no need to discuss it (emphases added, 52-54).

Apparently, "The Human Will to Live" cannot accept degrading pain, which coincides with the moment when pain affects a patient's mind and brain. Is pain ignored, or only seriously taken into consideration, when one's body is too depleted of its resources to any longer endure pain? Without promising answers, Lavery admits, at least indirectly, the meaningless of pain (a point of view that she shares with Vogel's play). Unlike the ambiguous end of this play's first act, Lavery proposes to conclude the second clearly. After June dies, her experience as a cancer patient will be dramatized with her friend, Joy, as the leading actress:

> JOY. (BIG BOTTLE OF MINERAL WATER, TALIKING TO A POTTED SMALL TREE…)
> Come here [i.e., addressing the potted tree].
> (SHE PULLS IT TO HER)
> This is the first conversation we've had sober. Me sober. So.
> (ENOURMOUS SHOUT) You killed yourself, you bastard!
> (SAME ENORMOUS SHOUT) And now, fucking June! So I'm not going to see her at the moment because I might just fucking kill her out of sheer imitation! […] You must have really wanted out, huh? I do understand, actually. I do. But. Dying's not just about you! (70-71)

Being confronted with a real problem (such as June's cancer) makes us realize that the praxis of pain is, paradoxically, a better way to define ourselves than by posing the same series of rhetorical and metaphysical questions about our (predestined) fate.

In this light, the novelty of Lavery's play is related to its Pirandellian quality. In her play, persons--disguised as actors--try to cheer June up, and by so doing to cheer themselves up. The play begins with Leah saying: "We are trying to cheer each other up because okay, June, this friend, this bitch, June's had breast cancer" (2). At the end of the play, "the bitch" will give her permission to let Joy theatrically interpret her cancer ordeal. In other words, what this dramatist achieves is to show June her "after," to which neither she (nor, by extension, we) has any access corporeally. She and her friends, together, meticulously work to make a believable and decent *mise-en-scène* of someone with cancer.

Their message may be that, even if cancer inexplicably will take June away from them, this illness need not take away their dignity as well. Retorts such as the following reinforce the idea that Lavery's characters are a group of people exhausted in their waiting for a cure for cancer and, instead, are now embarked upon a search for its dignified representation: "All the acting is impeccable" (25); when June pretends to sleep, and, according to Lavery's indications, "She looks dead. The others watch her" (50), Gash is not fully convinced of her performance's force and says, "Sorry. It's very undramatical" (50); and when Joy experiences that manic access of anger and yells at the potted tree, Gash retorts, "Drama-wise… Big anti-climax" (78).

When we do not understand things, events, and phenomena we try to define them. But when we do not understand the actual pain of the other, what do we do? Years after completing her book, *Illness as Metaphor* (1978), Sontag writes that its purpose "[w]as to calm imagination. […] Not to confer meaning […] but to deprive something of meaning. […] To regard cancer as if it were just a disease—a very serious one, but just a disease. Not a curse, not a punishment, not an embarrassment. Without 'meaning'" (14). Without meaning does not mean without significance--it does imply, however, without the excessive and thus obsessive interpretations we tend to give to a traumatic event, inevitably impoverishing it. Cancer is one of those words that can be defined only partially, since its sufferers experience its transitory manifestations. This is why Lavery's group of actors-friends-people meets daily to write, rehearse, and adjust the lines of a play with cancer as their *mise-en-abîme* subject.

While Vogel constructs her play on the alienation effect, pointing out the deliberate effect of a waltz danced alone, as noted earlier, Lavery's may appear to have a Pirandellian kinship. Describing Luigi Pirandello's

Six Characters in Search of One Author, David McDonald remarks that they are "[f]loating signifiers. Their story, identity begins and ends as a protowriting, a first writing, the mark and the trace of their *dasein*, their being-there. They have come in search of an author; they have come to be written out" (425). Lavery's actors are in search of a protoperformance of cancer, namely a first (for them) performance of this illness based on June's body's script that has been modifying drastically from one day and scene to another. In other words, as long as June's cancer constantly exhibits bodily and emotionally floating signs, her friends can best hope to stage a more or less impromptu protoperformance. McDonald also remarks that the six characters "[w]ant to gain their substance through performance, through enactment of their narration, through the act of speech and gesture [...]; to be not merely in the flesh or in the word but in the unity of both flesh and word" (426). Lavery's actors want to reach that symbiosis, too, between uttering arbitrary words and inconsistent flesh identity, between fluctuating form and modified matter.

Both Vogel and Lavery's plays pose a common concern; in the former, the character who has ATD/AIDS appears only fictively, while in the latter, the one with cancer sees herself transformed into a character; consequently, the climax of the plays is intentionally left unwritten, thus inviting the audience to participate. The last scenes of both plays are situated at what is actually some midpoint in the action, or they are perhaps set offstage, meaning in the middle of the audience. The way Carl is developed as a character may allude to a quiet, possibly passive, audience; on the other hand, Lavery's June functions as a *memento mori*, since she reminds us that we will become a memory, too.

Having briefly pointed this out, let me return to the main focus of both plays which is the *problem* of meaning, more specifically, the meaning of death as seen through AIDS and cancer. The playwrights reinvent the problem of meaning through a type of laughter that borders on tears; a similar reaction may result when we contemplate Edvard Munch's painting *The Scream* (1893). When we scream, the muscles of our faces are widely open, the diaphragm almost explodes in its attempt to release the anger locked within. Yet with what are we left once the scream subsides? What better opportunity to emphasize the contrapuntal technique than by saying that after anger, there must be joy; after screaming, laughter.

But, according to these playwrights, when angry, we should situate ourselves outside of ourselves so that we can properly acknowledge the laughable aspect of our anger and thus better balance our emotional lives. For example, to June, cancer "[i]s a 'neoplasm,' apparently... which has 'the ability to leave home and travel somewhere else.' Sort of Disease-Package-Tourist" (5). The first reading of cancer as a "neoplasm" arises as if June were looking over some notes taken without care in a doctor's office, or seeking out its definition from a dictionary. Immediately after the pseudo-scientific explanation given to "neoplasm," she adds her own, in a sort of laughter-and-crying kind of technique; without wanting, her cancer has made June a tourist in her own body.

This technique of combining laughter and tears is not new; as suggested by Vogel and Lavery, it has only been forgotten. In antiquity, theatrical festivals ended with a comedy that celebrated in disguise the foolishness of our actions and the cornucopia of our flaws. However, neither Vogel nor Lavery wants her play to be considered a comedy. They point out that sometimes, as June admits bluntly, "[s]hit happens [...] and the modus operandi for the happening shit, deal with it" (51). In this scenario, patients should rely on the wisdom learned from their enduring bodies. And most importantly of all, they should prepare the others for their departure, because these attendees will negotiate their deaths alone.

Yet this presents a dilemma. As Sontag admits, "Compassion is an unstable emotion." Although she places her emphasis on the other's participation, I take the liberty to use this succinct remark and ask: what happens with our compassion when our beloved die? Who comforts us? Undoubtedly, those who attend the dying are traversed by many self-doubts and fears *vis-à-vis* their own deaths. These individuals could become part of a solidarity-driven community by sharing their fears and agonies of loss. Because such a community does not (yet) exist, as proposed by Vogel and Lavery, another thing as sure as death is one's indulgence in self-prescribed treatments of laughter. In this light, everything could generate laughter, ranging from ridiculous gestures and actions, mixed-up reactions to illnesses' irrationality, meaninglessness, and death:

> GASH. This man dies. Goes to Hell. The Devil says... 'We have three rooms you can choose from. First room... everybody's standing up to their waists in shit. Second room... everybody's standing up to their necks in shit. Third room... everybody's standing knee-deep in shit, drinking cups of tea.' Devil says... 'Okay, which do you fancy?' The man says 'Well, if it's all the same to you... I'll take the third room.' Okay, says The Devil. Man goes in. The devil says, 'Okay. Tea break over. Back on your heads.' (62)

In *Endgame* (1958), Beckett inserts a joke within the dramatic text, too. His refers to an unsatisfied customer who repeatedly goes to his tailor to remind him that he is still waiting for his pair of trousers to be made. The tailor is lazy and it takes him a lot of time to complete this order. Dissatisfied, the customer reminds the tailor that God was able to complete His project in six days. The tailor replies shrewdly that God completed his project too fast, and hence His was full of errors. In other words, the tailor suggests that we are eager to have something quickly, disregarding its quality. In their own specific ways, Beckett, Vogel and Lavery teach their audience that one has to be prepared to deal with misfortunes, as they come unannounced in life, by rehearsing alternatives for surviving.

However, it seems in this case that the attendee cannot know whether he is being of any comfort to the dying loved one. This raises yet another question about the effect on the survivor. As inferred from these two plays, performing and narrating the pain of the other relies heavily on improvisation. Etymologically, this noun comes from the Latin verb *improvvisare*, which means "to see ahead." Without sufficient medical background, only few of us are indeed prepared to understand (at the level of discourse) the pain of the other. Even those medically educated cannot but improvise their reactions every now and then.

To make this point of view stronger, let us use as example the issue of rote memory; more specifically, when we learn a foreign language and its standard expressions.

> CARL. Répétez. En français. Where is my brother going? Où va mon frère? (Vogel, 33).

While we learn words, verbs, adjectives and the like in a foreign language, we cannot reach its uniqueness until we are confronted with real-life situations, namely when we speak sentences that make sense, and are not taken out of context and repeated mechanically. Similarly, when a beloved's body collapses, we have not had enough time to acknowledge it a breakable unity; therefore, we improvise our reactions because we try to make sense of what has become totally estranged from a once reassuring before.

At the level of narrating the pain of the other, we are additionally confronted with partial translations not only because there are physical and cognitive borders of empathy, but also because we must translate *ad-hoc* the jargon-ridden medical language. In Lavery's play, Joy shouts, "Dying is not about *you*!" (71, emphasis added). The deictic, personal pronoun "you" makes possible a dialogue, an embrace, a kiss, a touch, and, in a manner of speaking, contributes to the other's sense of identity. When "you" dies, a part of one's identity dies, too. There is no longer "you" and "me," but only "(impaired) me"; there remain memories, but they are retroactive and not active, part of the actual, certain domain. Put differently, "dying is *not* about you, but about us."

Therefore, when witnessing the other dying, we create thousands of thoughts, some healthy, others nightmarish. The Third Man, an obscure character in Vogel's play, believes: "There is a growing urge to fight the sickness of the body with the health of the body" (29). When the other is in pain, there is a constant, recurrent thought that s/he will return to what s/he used to be. We may also argue that this type of remark is a reflection of those transient speculations that transpire at night, in our dreams or obsessions.

Consequently, unlike Oedipus who poked his eyes to erase the memory of having slept with his own mother, the moment we saw a beloved dying, our most ardent wish relies on his body's willingness to restore its functions or, at least, die as dignified as possible. Towards the end of one's life, the refrain of hope is not chanted any longer, but almost invariably transformed into discovering strategies for coping with his eventual demise.

Silver's *Lost Train*

> In *Regarding the Pain of Others* (2004), Sontag remarks in her typically provocative manner that information about what is happening elsewhere, called 'news,' features conflict and violence –'If it bleeds, it leads' runs the venerable guideline of tabloids and twenty four hour headline news shows—to which response is compassion, or indignation, or titillation, or approval, as each misery heaves into view. (18)

She also reminds us that the "iconography of suffering" (*Regarding* 40) is not a new subject either in art or in daily life. However, she believes that we are attracted to a certain type of suffering that is typically collective (e.g., as resulting from belligerent, violent situations; in natural disasters, such as floods or earthquakes; hunger and promiscuity in third-developed countries; etc.). Carolyn Ellis asserts a similar concern; in her essay, "'There are Survivors': Telling a Story of Sudden Death," she remarks that it is about time for social scientists (and literary critics for that matter) to write about personal disasters and thereby make them public. As she argues,

> Although social scientists have written about disasters, their emphasis tends to be on the destruction of community, community behavior during disasters, community social order, and community mental health crisis intervention. But in airplane crashes, unlike most natural disasters, there is no community for survivors or families of victims. Passengers are

> strangers who have come from many regions, and survivors disperse quickly after accidents. (734)

Ellis wrote this essay in the aftermath of her brother's sudden death in an airplane crash. I think persons/patients who are in hospitals, persons/patients who can barely speak and/or walk, are in a frustratingly similar situation to the one described above. To make their pain more visible, this part of Chapter Four proposes a look at different aspects of hospitalization. More specifically, here I address the (limits of) bonding between persons/patients and their attendees. Although not related to each other, both Marisa Silver's short story and Ngozi Onwurah's film/documentary address the issue of intimacy, and how far we want or are prepared to assist the other's dealing with pain. Coincidentally, intimacy is discussed in both works through the lenses of filial relationship.

The first lines of Silver's "Night Train to Frankfurt" (2006) announce what will happen in a clinic in Frankfurt, Germany. On their way there by train, when thoughts do *not* fly as fast as a train's delirious pace, Helen (the narrator) confesses to us:

> They were going to boil Dorothy's [her mother] blood. Take it out, heat it, put it back in. The cancer would be gone. The treatment had a more formal-sounding name, thermosomething or other, a word that was both trustworthy (because you recognized the prefix) and lofty, so that you didn't question it. (77)

The passage is about choosing something, that is to say anything, when in crisis. It is about panicking; it is about no longer trusting traditional medicine that, faced with other cases of cancer, has not always been successful. Finally, it is about desperation.

Flipping through the clinic's advertising pamphlet, Helen is in a state of shock: could alternative medicine/treatments actually be more effective than traditional ones? And if they are better, then why aren't they advertised to the lay community? What is at stake here?

> [T]he pamphlet showed no images of the sick—a choice made [...] to deemphasize the questionable science behind the treatment. It would be impossible to look at a photo of someone as ill as, say, her fifty-seven years old mother and think that this faintly medieval idea, one that brought to mind leeches and exorcism, could succeed where modern medicine had failed, or, in Dorothy's case, where modern medicine had never been given the chance to go. (77)

One should not minimize the importance of modern medicine's accomplishments based solely on the above remark. However, one is tempted to say that sophisticated machines and (repeated) clinical tests are not the only remedies that persons/patients and their attendees look for when in crisis.

In this particular case, Dorothy did not opt for traditional medical treatments. We find out about her decision through Helen's voice, who is in fact the only narrator here. Silver's story concerns not only the persons'/patients' pain and suffering, exhaustion and fear. With the exception of the physical pain, which is not transferable, all the other effects apply to their attendees, too. Therefore, an illness questions not only one's strengths but also one's limits to endure undesired moments of exposed intimacy:

> Dorothy was as light and fragile as papier-mâché. Helen closed the bathroom door behind them, reached past her mother, and flipped up the metal toilet lid, then steadied Dorothy as she loosened her slacks and eased them down her lips. Dorothy had always been private with her body; Helen could not remember ever having seen her naked before the disease had turned her into a reluctant exhibitionist. (81)

Since illnesses are a complex of emotionally charged experiences, they can potentially bring us so close to our beloved that we may experience the blurring point of our identities. Like the *papier-mâché*, Dorothy's destitute body could now be "molded" (with the connotation manipulated), as if her body was already not human, but a marionette's. The end of Silver's story is ambiguous. Before they arrive at the clinic, Dorothy collapses on the train's floor. Since the author does not explain this moment, it could be interpreted either as her death or her possible recovery:

> Helen could see panic in Dorothy's eyes. […] 'This is it, Mom,' she said. 'This is the place. We just have to walk a few more steps and then we'll be there.' But just as she was about to put her arm around her mother, Dorothy drew herself up, somehow guided back to herself by her daughter's confident gesture and voice, and started forward on her own. (85)

However, based on what happens before this moment, Dorothy is probably not resourceful enough to regain her strength so easily, if not miraculously. This author refers not so much to the controversial act of healing, as to those "few more steps" that do not come effortlessly to us. In this passage, Silver emphasizes our ritual of pleading for a few more steps, a few more days, or a few more minutes to figure out a problem and escape this existential maze.

Onwurah's Perspective

This feeling of being entrapped is presented in Onwurah's *The Body Beautiful* (2001), too. The key passage here happens in a sauna, where mother, daughter and other (healthy) women share a moment of awkward intimacy. The director's emphasis is placed not on how strangers share a sauna, but rather on the division between the healthy and the ill body, as the women in the sauna seem to scrutinize the ill woman's body with their too intrusive gaze.

Madge, the person/patient in this film, discovers she has breast cancer while pregnant. Immediately after delivery, her body undergoes a surgical intervention, which consists in the removal of her cancerous breast. Because of this surgery, she cannot breast feed her newborn. She says, "[a] child screams after my milk," but she is powerless. She feels cheated. Years after her mastectomy, her body challenges her patience again when she is diagnosed with a severe form of rheumatism. She thinks pain "crucified her" and it "chopped inside" of her. She perceives herself as unjustly deformed. Still, she craves for an embrace, simple, without any meaning, which she envisions to "[s]mooth up the deformities."

Now that we have learnt about this woman's suffering, we wait to see the director's response to her ordeal. Onwurah proposes a polemical approach to sympathy. Madge's daughter is young, healthy, and has a perfect, robust body. She works as a model. Her body's stamina is suggested through its every pore. Thus the daughter cannot fully feel what her mother feels. After the frame in the sauna, the daughter, sitting naked in front of a mirror, presses hard on her breasts; she wants to imagine what it would feel like to have no breasts at all. Then she says that the project does not make any sense: "[i]t is like closing one's eyes to feel like a blind [person] only to open them again." Onwurah's message is that the other's pain is not sharable, and that trying to imagine what the other feels is an unnecessary added discomfort.

But if there are limits to cognitive empathy, there should not be barriers to emotional empathy, to touching and holding the other in pain. This is in fact how Onwurah ends his short movie, where mother-and-daughter rest in bed naked, holding each other. The embrace creates the illusion of one body that has intertwined tissues of young and healthy, sick and old, positive and negative. This last frame is an elaborated tribute to Plato's ideal couple; in *Symposium*, Aristophanes structures his rhetorical arguments on the nature of Love focusing on a lost time when couples were literally whole, and not separated/divided. Through their embrace--a short kiss of their bodily epidermis--Onwurah's mother and daughter unite love with pain, hope with loss.

Almodóvar's *Talk to Her*

The last section of this chapter focuses on Almodóvar's film *Talk to Her* (2002), which concerns how we are affected and transformed by loneliness. The director intrigues us by creating an interesting conflation between devotion and la *soledad* ("loneliness"). However, one feels trapped in this film's scenario and uncertain about where to place one's sympathies. With Alicia, who has been comatose for four years, but whom Benigno (her nurse) has watched constantly with devotion? Or with the latter since his devotion and care are not consciously reciprocated? And yes, I did write "conscious," because Almodóvar insists that our body's force of communication is preeminently non-verbal. He begins his film melodically by shooting a ballet scene, which, according to Marsha Kinder, "[p]ositioned as a prologue to the film, the 'Café Müller' sequence enables us to read the movements of the characters in the subsequent story as a dance" (D'Lugo 108). There is something else about that pro-logue when the dance is expressing only the pulse of desire and the choreography of touching. At the end of the ballet, the audience may feel enraptured, but also physically exhausted since their muscles are a little bit atrophied from the prolonged sitting and inactivity. Carl Plantinga argues that, "Since our minds are 'modular' and capable of parallel processing, we can monitor another's emotions while doing other things, for example carrying on a conversation or following a narrative" (243). Can we monitor the other's emotions when s/he is in pain? In such situations, are we capable of a meaningful "parallel processing"? Can we simultaneously be "there" for them and "here" for us? Or is this an example of how we impose distance and thus separation?

Marco, the second protagonist in this film, is skeptical that communication could be effectively achieved through non-verbal acts. As a reporter, his understanding of life relies heavily on words. Marco's girlfriend, Lydia, is gored during a *corrida*, after which she becomes comatose. Unlike Benigno, Marco cannot either touch or talk to Lydia, and he gives the impression he finds Benigno's advice and its practice unhealthy and unethical.

Benigno speaks and dedicates his passion to a woman who is comatose, and for whom Time stands still. He is attracted to Alicia, indisputably, but at the same time he is infatuated with the stillness of her body, which reciprocates the paralysis of time, before it will be reacquainted with its small divisions of hours, minutes, seconds, and fractions of seconds. Almodóvar alludes to a beneficial stillness. According to Gregory Flaxman, the "cinema of inaction" is characterized by "waiting and exhaustion; the image does not extend to new spaces but 'intends,' involuting into the mind, opening up to a whole new sense of mental duration (durée), an involution into psychic states" (6). "Intends" is an interesting word choice. Henri Lafebvre in *The Production of Space* (1991) argues that

> [t]he living being constitutes itself from the onset as an internal space. Very early on, in phylogenesis as in the genesis of the individual organism, and indentation forms in the cellular mass. A cavity gradually takes shape, [...] which is filled with fluids. [...] The space thus produced will eventually take on the most varied forms, [...] beginning with an initial stage at which it has the form of what the embryologists call a 'gastrula.' (176)

Dictionaries define gastrula as resulting from the invagination of bastula. Almodóvar plays a lot with the principle of invagination, interiority, cavities and other inward, concave spaces, because he wants to return to the mind of the body--the supreme interior cavity--which is fraught with concepts, ideas, and above all emotions. His return to the mind of the body is accomplished through love and sacrifice, and not through exhausting acts of meditation.

Interestingly, there is a good deal of intertextuality between Almodóvar's film and Harold Pinter's play *A Kind of Alaska* (1982), where the latter envisions a tough perspective on the theme of sympathy. Like Alicia, Deborah has been comatose. In the preface to the play, Pinter reports that during "the winter of 1916-1917" (3) an epidemic of *encephalitis lethargica* left many victims among Europeans, keeping them asleep until a drug was invented and tested successfully fifty years later. Unlike Alicia, Deborah remains in a state of disbelief *vis-à-vis* her prolonged traumatic experience. When Pauline starts to talk to her sister, the latter takes the former to be her aunt. When Pauline asks Deborah, "Do you remember me?" (27), what intrigues Deborah is not the question per se, but seeing this stranger addressing her a question to which she should know the answer. Yet because she was comatose for twenty-nine years, she does not know, let alone remember, Pauline as a woman; she remembers her only as a girl. But Pinter does not focus on Deborah, whose frustration and state of disbelief are highly understandable. Instead, Pinter suggests that pain is frustrating, and that the choral figures' patience eventually empties itself:

> HORNBY. I have been your doctor for many years. This is your sister. Your father is blind. [...] Your mother is dead. [...] You see, you have been nowhere, absent, indifferent. *It is we who have suffered.* (34, emphasis added)

Hornby's "It is we who have suffered," placed next to Lavery's Joy's remark, "Dying's not just about you!" brings to our attention the divide between persons/patients and their attendees. If dying is not just about you, then, by the same token, suffering is not only about you either. Or, as proposed by Pinter and Lavery, if you are in pain, then *I* am in pain. Therefore, an illness creates copulative spaces between persons/patients and their attendees. But underneath these spaces, there are disjunctive spaces dominated by fear, anxiety and futility, whose existence we feel ashamed to admit.

In this light, Almodóvar's Alicia may appear too composed when she casually enjoys a ballet show in the last frame of the film. On the other hand, Pinter's Deborah refuses to believe what happened to her, preferring instead to live in her past:

> DEBORAH. She [Pauline] is a widower. She does not go to her ballet classes anymore. Mummy and Daddy and Estelle are on a world cruise. They've stopped in Bangkok. It will be my birthday soon. I think I have the matter in proportion. (40)

Yet, from another perspective, she has the matter in portions, not proportion. The much awaited moment of her waking up does not bring comfort to those who have already suffered for twenty nine years. The very last lines of the play are composed using a contrapuntal technique. Only Deborah's first statement is actually accurate. Pauline is a widower, as Hornby admitted earlier in the play. He said he spent more time sitting next to Deborah than living with his (real) wife. However, once Deborah states that her sister is a widower, her contact with reality is lost again. Apart from this character's refusal to accept as past/gone a life she never had for twenty nine years, this passage addresses the negative side-effects of our rote memory. Deborah's incoherent speech may allude to the type of knowledge we learn by heart, and not by conviction, and then reproduce it *ad litteram* (i.e., Pauline is a widower. But that does not mean anything to Deborah).

Furthermore, Pinter admits how little we know ourselves once chronic suffering mocks our carefully protected identities. As the playwright declared recently, "When we look into a mirror we think the image that confronts us is accurate. But move a millimeter and the image changes. We are actually looking at a never-ending range of reflections" (818). We should not assume that these reflections are about hope or comfort, for there is none. In his typically polemical manner, Pinter continues his train of thought: "But sometimes a writer has to smash the mirror—for it is on the other side of the mirror that the truth stares at us" (818). When he made this remark, upon his acceptance of the Nobel Prize in literature, he did not give any examples, advice, or solutions. I quoted his reflection solely because an illness makes us stare at ourselves, too. We stare because we are incapable of finding solutions to suffering that could offer comfort for *both* the person/patient and his/her attendees.

Then, our bodies get exhausted when trapped in too many discourses and excessive worries. As example, let us focus on the short, black-and-white film-within-the-film entitled *Amante Menguante*, or "The Shrinking Lover" (from Almodóvar's *Talk to Her*). This clip is a eulogy to passion, viscera, and togetherness. Although the title of the Spanish director's film appears to be an order/advice, "Talk to Her," decoded from the silent clip inserted in the film, we realize that this film is about movement, dance, embrace, touch, and other non-communicative means that have a long lasting force to express our feelings for a beloved one. "Talk to Her" becomes "Touch her" (and by extension him). When we talk too much, we lose ourselves, just as a stone loses its mass when it is chafed constantly. In addition, when we talk on the other's behalf, we may unintentionally violate his/her experience.

Consequently, as proposed by Almodóvar, his film is about the supreme form of devotion. Benigno, who committed the unthinkable (raping Alicia), did that in a desperate, final attempt to possess the (absent) body of his passion. In prison, he refuses to live a life in which he cannot touch, talk to, or see Alicia; therefore, he kills himself. To better understand Benigno's gesture, or sacrifice, we should raise these questions: Is compassion an(other) unsynchronized emotion? If pain is not physically sharable, but it makes both the person/patient and his/her attendees suffer, what becomes of compassion once our beloved die? It appears, then, that compassion must be thought of as one strong emotion that redefines our negotiations with life's unpredictable, delicate moments. As the root of the word indicates, compassion is to suffer together.

Conclusion

This chapter, however, has arrived at a different vision from the choruses that concluded the ancient Greek plays with their moralizing speeches. Somehow, the other's pain erases the prefix "com-" from compassion, thus becoming "passion." In other words, one turns the other's experience of pain into a way of reflecting upon one's own meaning. This vision may help explain why the sections of this chapter seem not to cohere perfectly with one another, as if they were parts of an incomplete journal. I purposely compose them in a vignette-like style, because to speak on the other's behalf is a process that requires a lot of individualization, and thus self-absorption of pain.

The challenge is whether or not we could truly sooth the other's pain. Herein lies a great divide or separation, if you will. When we look far forward at the horizon so that everything seems to be seamless, the more we walk, the clearer and more distinct things crystallize into view. The same vision is expressed by Erica Berger's photograph *Last Days* (1992), the visual epigraph of this chapter. The attendee does not show his face since he has collapsed emotionally from the pain of the other. His proximity to the patient is evidently irrefutable. However, their visual foci do not intersect with each other; then, no matter how much closeness the experience of pain has brought them, their space--like their identity--reveals the significant degree of their inescapable separation. Put differently, confronted with and changed by the drama of the other's pain, the attendee develops an "absorption-in-delicate-moments" *syndrome* where he cannot totally understand his beloved's pain, nor, unfortunately, negotiate its fluctuating meaning.

CHAPTER FIVE

ARS MORIENDI

> To lose the place and time of experience is to lose the skin, to disclose the skin, not as a scene, but as the absence of a scene.
> (Connor 50)

> Hippocrates opens his description of *The Physician* with the observation that 'the dignity of a physician requires that he should look healthy; [...] for the common crowd consider those who are not of excellent bodily condition to be unable to take care of others.'
> (Gilman 52)

> Just as Dionysian terror is necessary to tragedy according to Nietzsche, Medusan fascination is necessary to the image according to Lacan.
> (Foster 280)

Sander L. Gilman, in his book *Picturing Health and Illness: Images of Identity and Difference* (1995), argues that during the 19th -century the healthy norm perceived as ugly not only those who were deformed, but also those who were ill, ageing, and/or experienced different bodily "loss of function" (53). He suggests that "The healthy are at the baseline for any definition of the acceptable human being, as if the changes of the body, labeled as illness or ageing or disability, were foreign to the definition of the 'real' human body" (53). In the 19th-century, how much was medicine responsible for defining ugly as ill, deformed, and getting old, versus beautiful as healthy, and then, for the sake of the community's health, firmly promoting these ideas? Furthermore, with the rise of photographic art, medicine was able to manipulate and control these ideas even more efficiently. According to Lupton, "The new technology of photography that developed from the mid-19th century became a valuable strategy in the documentation of patterns of disease and illness, and the construction of the sites of dirtiness and contagion" (*The Imperative* 30). She does not elaborate the notion of "patterns of disease and illness," although one could believe them to be part of a larger campaign of educating and controlling people; these patterns are not exclusively medical and/or pathological in nature, but they are also socio-politically constructed. As Lupton notes further, "[p]hotography was an important tool in the documenting of the urban spaces that required surveillance" (*The Imperative* 30). Put differently, photography is an art of recording personal memories, as well as a mechanical technique that captures one episode (after another) in the evolution of the modern human body--both in its physical and political dimensions.

Undoubtedly, photographs have a hidden dimension that derives from their unspoken invitation to be touched, and that in return generates the chiasmatic delineation between perceiver and perceived, beholder and beheld. Or, as Steven Connor describes the tangibility of photographs,

> This sheen signifies the magical preciousness that we wish the photograph to retain, giving the eye notice that it is a tangible thing which can never be encompassed simply by looking. The gloss is the ideal skin, flesh transfigured, but the identity of that skin is what seems to guarantee its quality of tenderness, that word that signifies both the quality of something touched and the manner of our touching. (37)

Our bodies and skins share something in common: just as our skin covers our bodies, not revealing what lies beneath, so our skins are covered by make-up, plastic surgery, lotions, or clothes, thus, again, partly presenting their "stories."

In this chapter, I focus on the skin's narrative as it exposes its story when photographed. David Wojnarowicz, William Yang, and Jo Spence capture the body in three different hypostases: dead, dying, and ill. Wojnarowicz rarely exposes himself as the object of his art; when he does, he focuses primarily on his face, and not on his entire body. One of the subjects of his artwork is a rather disturbing hypostatization of his mentor and partner, Peter Hujar, lying dead in bed. Yang takes photos of his good friend, Allan, who is dying of AIDS. Finally, Spence reveals her nude body along with its signs of cancer. What interests me here is to discuss/approach the photographic art not from its scopophilic angle, that is, not from its perverse and pleasurable voyeuristic angle, but to analyze it side-by-side with the double referentiality attributed to our bodies. As Merleau-Ponty suggests,

> [w]hen I touch my right hand with my left, my right hand, as an object, has the strange property of being able to feel too. [...] I can identify the hand touched as the same one which will in a moment be touching. [...] [The body] tries to touch itself while being touched, and initiates 'a kind of reflection' which is sufficient to distinguish it from objects of which I can indeed say that they 'touch' my body, but only when it is inert, and

> therefore without ever catching it unawares in its exploring function. (106-07)

The photographs that are included in this chapter reveal the inert dimension of the body in pain. Here I do not refer to the incapacity of this art to retain some of the persons'/patients' pain and bodily discomfort; instead, I argue that these photographs become emblematic for what Leder spoke of as "the remaining body." To understand the functioning, or maybe I should say malfunctioning, of such a body, Leder develops his arguments from

> [t]he modes of un-readiness-to-hand that Heidegger describes in relation to the tool. At times of illness one may experience one's body as more or less 'unusable.' It no longer can do what once it could. [...] The body that remains, as Heidegger writes of equipment, 'reveals itself as something just present-at-hand and no more, which cannot be budged without the thing that is missing. (84)

The three artists' works analyzed in this chapter offer variations of the "body that remains," and, as we shall see, of the body that gradually did *not* remain. Through their work, Wojnarowicz, Yang, and Spence approach visually the theme of the *ars* moriendi of the entropic body in pain. The photos presented here constitute a delicate means that reflects the inconsistencies presented in the act of remembering the details of how one person's body traveled through time, how it used to be and look. Because these photos refuse to connect their subjects to their past, I suggest that the photos reach a type of quietude attributed to the dying body. As a consequence, when the bodily "loss of function" becomes more and more noticeable, while the body becomes less and less visible, I argue that the space captured in the photos loses its isotropic quality where "[i]ts relations are considered symmetrical" (Grosz, *Space* 96), meaning identical in all directions, and it becomes anisotropic, a quality which Grosz describes as belonging exclusively to time, where "[t]he relations of before and after are regulated" (Grosz, *Space* 96). In these photos, only the outside, chronometric time is regulated in its units of before and after. The photos shown in this chapter allude to a moment that may not have continuity in time, either in its backwards or forwards directions. I read these photos as being spatially and temporally discontinued by the pain affecting and disrupting the continuity of their subjects' lives, who are confronted with the immanence of their death.

David Wojnarowicz

In 1869, homosexuality became classified as a medical problem in Western societies, homosexuals being considered medical objects for the study of abnormal behavior. The modern homophile movement did not emerge until after the Second World War, when homosexuals began their effort to eliminate such notions as: "medical object," "mentally ill person" and, especially, "the sinner." During this stage, which occurred from the 1950s to the 1960s, homosexuals hoped that society would eventually accept their personal sexual choice. Yet understanding that without a social movement, spokespersons, and sympathizers, homosexuals are unlikely ever to be accepted, or at least tolerated by the majority, they began to organize themselves better. One of the most important of these organizations was the Gay Liberation Front, which was born out of a violent confrontation between gay bar patrons and police at the Stonewall Tavern in Greenwich Village, New York, in the summer of 1969. That was a major step forward in their liberation, which revealed the conflict between the gay subculture and the larger American society. Furthermore, in the late 1970s, some church denominations began to welcome gay members and even clergy. This represented a breakthrough in their movement since the Anti-Gay Movement's rhetoric relied heavily on moral aspects derived from secular, religious discourses. In this context, the future for homosexuals looked bright, full of hope, since the old prejudices against them seemed to be perceived as less threatening to society.

Then, in the late 1980, AIDS was officially diagnosed, first limited to sexual fast trackers in San Francisco and New York, but quickly spreading within the larger gay community. By 1985, the mass-media were claiming that AIDS was an exclusively "gay plague," which, via drug addicts, threatened the larger strata of society. Because of the misinterpretation of AIDS, the Anti-Gay Movement began to win more supporters to their side.

The artworks of David Wojnarowicz have contributed substantially to making the gay community visible once again. After a traumatic and abused childhood and adolescence, he sought comfort in admiring, contemplating, being in the presence of art, and eventually making art as a gifted visual artist and a refined writer. His feelings of a traumatized adolescence unfortunately did not end once he reached his mature years. On the contrary, he seemed to have been simply betrayed by an unfair destiny (he died of complications of AIDS at age thirty-seven). In an entry in his memoir, *Close to the Knives: A Memoir of Disintegration* (1991), tired and disappointed with others' hypocrisy and polite pretense, he writes: "I wish I could get a selective lobotomy and rearrange my senses so that all I could see is the color blue; no images, or forms, no sounds or sensations" (7).

Among Wojnarowicz's first notable artistic attempts were the photographs in the "Rimbaud series." Arthur Rimbaud was a 19th -century French Symbolist poet, who like many symbolists, emphasized the synesthetic quality of our senses, their intensified perception and function when used together. In his works, however, Wojnarowicz focuses on the opposite: the non-sense/irrational side of our senses, the abuse of reason, and the inequality of freedom and human rights. He believes we are born into a "[p]re-invented existence" in which "[w]e either adapt to it or end up dead or barely surviving" (qtd. in Rizk 45). During his Rimbaud period, the tone is still sensual and not menacing, and he employs the art of photography combined with collage; namely, on some photos of himself or his friend (capturing them naked, on deserted seashores, or in the crowds of the New York subway) he invariably pastes the same image of Rimbaud's face, as if he or his friend did not have any identity, and, in order to be recognized by the crowds, *had* to wear the mask of a famous man. A few years later, Sontag would say that "All the debunking of the Cartesian separation of mind and body by modern philosophy and modern science has not reduced by one iota this culture's conviction of the separation of face and body" (*AIDS* 39). Intuitively, Wojnarowicz may have felt that, too. During that phase he discovers his artistic signature--the collage--which he believes is best suited to describe his feelings because of individuals' "[e]xperience to society's pre-existing contexts" (qtd. in Rizk 50). The collage does not respect the laws of the linearity of the arrow of time, and it makes clash together in one image different spatial and temporal units. Hence, a collage is plurivocal.

For example, in *Fuck You Faggot Fucker* (1984), Wojnarowicz frames the central couple with four black-and-white photos of naked men. The central couple depicts two men, from torso up, embracing and kissing each other; their bodies are malleable and retain whatever shape one molds them into. In this case, it is not their shape that is altered; they have lost their skin onto which a map of the globe has been grafted. On closer examination, we notice that the map itself has been under an external pressure, since it is represented as if it were torn apart from its larger image, thus probably alluding to the inequality of freedom and the human right to privacy that exist globally. Below the central image, there is graffiti with two men caught in their copulating act. The graffiti reads "Faggot," "Faggot fucker," and "Cats shit," once again pejoratively addressing the homosexuals, and, what is worse, placing them in an offending scatological category. It is by far the most extreme denigration of the body, as opposed to the one presented in the works of the 16th-century French writer, François Rabelais, where the body enjoyed and treasured all its orifices. In fact, some modern medical discourses have created this dichotomy between the classical body (self-contained and culturally bound, hence sophisticated) and the grotesque body which is "close to nature in its uncontrolled state" (Lupton, *The Imperative* 8). In this light, homosexuals may be perceived as engaging in grotesque, unhealthy, and thus uncontrolled acts.

The visual quality of this work aside, what catches my attention is actually its aggressive title, *Fuck You Faggot Fucker*. Within the rhetoric of unrepresented groups there exists an immense amount of self-pity, making them feel like scapegoats and, in most cases, experiencing self-hatred. To redress this problem, in their speeches, they employ a "perspective by incongruity" technique, where hyperbole and/or overstatements show sarcasm, possess a caustic tone, and are used for a "wake-up call." In his 2004 speech, *The Tragedy of Today's Gays*, given five days before the reelection of President George W. Bush, the playwright and activist Larry Kramer says, "This past week almost sixty million of our so-called fellow Americans voted against us. Indeed, twenty three percent of self-identified gay people voted against us, too. That I can't figure. Please note that a huge portion of the population of the United States hates us" (36-37). Years earlier, in Kramer's play *The Normal Heart* (1985), the doctor informs her HIV positive patient that the doctors "[t]ry to stay out of anything that smells political, and this smells. Bad" (34). Whether HIV positive or not, Kramer believes that all homosexuals are still perceived as pariah, and hence rejected and hated.

To make this point clearer, namely that homosexuals think of themselves as an ostracized minority, let us examine Wojnarowicz's work, *Untitled* (ca. 1983), where the artist is caught in motion by the camera; he attempts to fly, but he will fail. Interestingly, Leonard Shlain believes the verbs "to fail" and "to fall" have a common descendant from the Latin verb *fallare*, "to fail, to deceive" (308). Metaphorically speaking, as human beings we are afraid of falling: "Homo sapiens is first and foremost a primate. Although none of us flies among the treetops anymore, we still retain buried deep within our archaic collective memory an atavistic fear of falling" (Shlain 308). Wojnarowicz deliberately places himself next to a representation of an extinct creature--the pterodactyl--suggesting the homosexuals' official denial of citizenship, and hence their exclusion could be understood as metaphorical extinction.

However, to admit extinction actually means to feel guilty and accept your "deserved" punishment. In a later work, *Silence = Death* (1990), a still from a video, Wojnarowicz creates the impression that he has sown his lips. The illusion could pass as stitches, implying a profound cut that could not be taken care of with only a bandage (which would once again conceal rather than reveal the severity of the wound). The sewing of the lips is also erotically charged, but if the lips are so violently closed, then this act may be read as abdication of one's own private pleasures.

The sewing of his lips urges an "either"/ "or" reading. If Wojnarowicz is not permitted the chance to speak up, then he will stop talking (and eating) for good. He believes that it is time the lay community heard the homosexuals' side of the story; that AIDS is not an affliction generated by homosexuals; that, on repeated occasions, the government spent far too little money researching this illness; that the research conducted so far has proved beyond doubt that AIDS is an illness with uncertain etiology; that the lay community should be correctly, and not partly, informed, and thus properly educated.

In his speech, Kramer points to some of these ardent issues, too. As he says,

> I do not understand why some of you believe that because we have drugs that deal with the virus more or less effectively that it is worth the gamble to have unprotected sex. [...] These drugs are not easy to take. There are many side effects. [...] Many of the meds we are now taking are new meds and were approved quickly and side effects have a sneaky way of showing up after FDA approval, not before. I recently discovered that I was taking an FDA-approved dose of Viread that has turned out to be five times the amount I actually need. (44-45).

Both Wojnarowicz's *Silence = Death* and Kramer's speech bring into discussion an Enlightenment maxim, more specifically *Aude Sapere*, "[d]are to know by daring to think for yourself. And this need to think for ourselves is likewise where we find a particularly modern democratic conception of equality that also implies freedom" (Shusterman 95). One of the issues that should be known by the public is that HIV is a retrovirus, meaning its effects appear only years later; because of that, the virus may be transmitted to others unintentionally. Wojnarowicz considers his beloved partner's death as caused, at least in part, by this *abused* public misinformation. In his *Untitled [Hujar Dead]* (1988-89), Wojnarowicz shoots the last images of his beloved, lying motionless in bed. The temporal inertia experienced by many persons/patients on their deathbed is here confronted by these morbid, petrifying, yet solemn images of Hujar's corpse.

How do these images alter our perception of photographs? We usually view photographs as part of an album, and not the point terminus in representing one's successive stages of embodiment. In his essay, "Understanding a Photograph," John Berger writes,

> What varies is the intensity with which we are made aware of the poles of absence and presence. Between these two poles photography finds its proper meaning. (The most popular use of the photograph is a memento of the absent.) […] The language in which a photograph deals is the language of events. All its references are external to itself. Hence the continuum. (293)

Wojnarowicz's images of Hujar dead show his lover's last phase of embodiment, and they represent an evocative variation upon the classical theme of *memento mori*. The artist does not emphasize the idea of "remember that you will die," presenting images of skeletons, as this theme has been interpreted usually to reinforce its moral implications along with providing a reminder of our bodies' physical limitations, but he focuses instead on *how* one dies. In his phenomenological philosophy, Merleau-Ponty insisted that we have a body in the sense of how we know or manage to live it. Could we go as far as Wojnarowicz seems to suggest by posing this question: does how we die somehow reflect, *post-mortem*, how we lived our bodies? To understand this question's implications, one should return to Wojnarowicz's image. The photos showing close-ups of Hujar's half-open eyes, hands, palms, and face are framed by images of dollar bills and semen. The overlaid text employed the stencil technique, which is known to have "[t]he eye-catching characteristic of seeming both handmade and machine processed" (Cameron 7). Some excerpts from the text read: "'If I had a dollar to spend for healthcare I'd rather spend it on a baby or innocent person with some defect or illness not of their own responsibility; not some person with AIDS…', says the healthcare official on national television" or, "'If you want to stop AIDS shoot the queers… says the governor of texas on the radio and his press secretary later claims that the governor was only joking and didn't know the microphone was turned on," or "as each T-cell disappears from my body it's replaced by ten pounds of pressure ten pounds of rage […] it's been murder on a daily basis for eight count them eight long years" (60). The text does not make any specific reference to Hujar's death, since Wojnarowicz did not want to focus on *one* particular death from AIDS, but on innumerable unknown others.

Therefore, the text expresses his belief that "silence equals death," and that one should dare to know and seek alternative sources of information, since the mainstream may rarely be accurate. He believes that the society in which he lives refuses to "deal with mortality" (*Close* 7), and he blames this failure on a lack of shared intimacy. As he writes,

> To make the *private* into something public is an act that has terrific repercussions in the pre-invented world. [...] One of the first steps in making the private grief public is the ritual of memorials. [After attending some of them, he writes,] [t]he memorial had little reverberation outside the room it was held in. a TV commercial for handiwipes had a higher impact on the society at large. (*Close* 10-11)

One such public artwork created by the NAMES project consisted in a huge quilt that commemorated those who died of AIDS. The quilt was displayed in Washington, D.C. in 1996, facing the Capitol Building (The location was chosen carefully; there, in 1963, Martin Luther King Jr. had presented his legendary "I Have a Dream" speech). About the quilt, Joann P. Krieg notes that "The most unusual memorial to those who have died of AIDS has come not from the world of literature or drama but from handcraft—specifically, a type of needlework that is quintessential American, quilting" (142). Yet, that project and other fundraising efforts dedicated to AIDS still seem insufficient compared to the needs of this complex of illnesses.

Consequently, I wonder if AIDS has, in fact, become "banalized," just as Sontag pointed out earlier when she claimed that "AIDS has banalized cancer" (*AIDS* 44). Although this illness has taught us to be more cautious in our sexual relationships, it is not the main priority in medicine today, since new issues have recently arisen to shake our confidence in a body that, despite all the medical and technical revolutions, could hardly be kept intact, away from physical breakages and/or severe illnesses. Kramer suggests that today "[s]cientists now warn that even partners who are both uninfected should practice safe sex. As I understand it, more and more new viruses and mutant viruses and partial viruses that are not understood are floating around" (50). The hygienic movement, as part of a healthy lifestyle which emerged in the 19th -century, seems to be again questioned. The term plague, employed when we first described this illness (AIDS as a "gay plague"), is not used anymore. Yet, people are still diagnosed with (and die because of) AIDS. But the danger in thinking about AIDS as a plague is that such reasoning would somehow make us mentally revisit primitive times (i.e., not scientific times) when plagues destroyed cities and communities. Yet in refusing to conceive of AIDS as having a similar impact on our society, we may downplay the continued gravity of this illness. Wojnarowicz's artworks remind us forcefully that AIDS is not an illness that exclusively affects and/or is generated within the homosexual community, and hence our bigotry in regarding their intimacy as being abnormal must also be questioned.

William Yang

In the section of his work dedicated to AIDS, Gilman discusses only a collection of posters that have circulated in mass-media, which he researched at the National Library of Medicine at Bethesda, Maryland. Gilman thinks these posters function as the "still images of illness" (174). In other words, he believes these posters may have had an impact on the lay community, although not the intensified, urgent one, as he would have hoped. As he remarks about these posters, "What strikes a viewer [...] is the virtual absence of images of the diseased body. [...] Death certainly threatens, but dying itself is absent. While death is present, *dying is absent*" (118, emphasis added). Because Gilman did not include a single photo of a person/patient dying of AIDS--although he understood this lack--in this part of the chapter I juxtapose one of the posters from his book with William Yang's photos taken of his dying friend, Allan, from his project entitled *Sadness. A Monologue with Slides* (1988). Here I am interested in the impact of Allan's increasingly emaciated body versus the static, almost ineffective quality of the poster in order to consider the idea according to which "AIDS victims are living sculptures. [...] Both subject and object of art [...] they combine with their disease to overcome the narcissism of human consciousness. [...] It is an art of continuous transformation of subject into object and object into subject" (Siebers 220-21).

The art of the artists themselves affected with (or about others who have) AIDS is disjointed from the artistic community for various reasons: they feel estranged from their bodies' unpredictable fluctuations; they live in the present progressive of their bodies, which are gradually debilitated by their illness; and they feel detached from a past with which they can no longer reconnect. Therefore, I think that a new type of interiority emerges; however, because we live in a culture that exhibits a passion for instantaneity and consumption, this new interiority must be created *and* consumed immediately, without allowing time to reflect upon it. The photos (especially the mechanical ones) attempt to recapture something from this lost perception of time, and those taken of Allan serve as a good example.

William Yang is an Australian artist with Chinese parentage. The images presented in this chapter originally appeared in print in Thomas W.

Sokolowski's and Rosalind Solomon's collection of essays entitled *Portraits in the Time of AIDS* (1988). According to the editors, Yang presented them as "monologues with slide projection in the theatre" (34), because the main actor of this one-man show is dying of AIDS. Yang's work consists of seventeen slides with short texts written underneath them. In an attempt to respect the body that is dying, the texts are not recited, but the readers/spectators read them subvocally. The brilliance of this piece resides in its hushed tone, which parallels the act of dying when the persons'/patients' body and mind become more and more tacit and lifeless.

From one photo to another, and from one text to another, we discover Allan, although we never quite get to know him. The minitexts relate Allan's story: how he was hospitalized at St. Vincent's, known as "the AIDS ward" (35); how he decided to return home, into a studio shared with a dealer; how AIDS first attacked his lungs, and so he had to keep next to him "a large cylinder of oxygen as he was often out of breath" (37); how AIDS then affected his sight, and he developed a condition known as "CytoMegalo Virus—C.M.V. Retinctus" that gradually "destroyed the retina" of his eyes (39); how he decided "to go off medication" "(46); and, how, finally "he went into a coma. I saw a nurse give him a glass of water but the water just ran out of his mouth" (50). This is the information Yang offers us about Allan and that intrigues and saddens me profoundly because I never have the opportunity to know him as a person. Who was Allan before his diagnosis of AIDS? What did he do? What were his passions? All these questions are put in the past tense, and they will remain unanswered. Hence, Allan becomes the epitome of the body dying.

To look at these photos time and time again is to be reminded of Albert Einstein's vision of the passenger trapped in the train running with the speed of light. That passenger could not sense all that was happening in the train, and especially outside of it, because time moves in its cosmic, non-human, slippery dimension, and thus sensation could not profusely permeate his body. Juxtaposing Einstein's vision with Allan's decaying body, I read the latter's body as if it were coiled up inside his mind just like a snail covers a part of its body under its hard shell. The photos are presented rapidly with no entr-acte in between; in a matter of minutes, time and space seem to collapse. There is no time for a prolonged reminiscence of one's spent life, as--by contrast--is provided to Krapp, the only character in Beckett's play *Krapp's Last Tape* (1958). Krapp is not dying; something in his past died, and here lies the most crucial difference between him and Allan. Allan is dying *now*, and he does not have time to remember his life. He barely has time to feel his body, a touch, or a kiss on his face, which seems to Yang "to have caved in" (47).

Through this work, not only does Yang capture the disturbing moments of a friend dying, but he also touches on the epidermis of despair. This epidermis is both endotopic and exotopic, meaning that it starts within the person/patient and then it radiates/extends to his/her relatives/friends. How is the viewer positioned *vis-à-vis* a dying body? In his book, *The Birth of the Clinic: An Archeology of Medical Perception* (1973), Foucault, referring to the process of integrating meaning into the act of seeing, writes:

> To describe is to follow the ordering of the manifestations, but it is also to follow the ordering of the intelligible sequence of their genesis. It is to see and to know at the same time, because by saying what one sees, one integrates it spontaneously into knowledge. (114)

Put differently, to see is to know that you feel. Consequently, as Foucault argues, "[t]hings seen can be heard at last, and heard solely by virtue of the fact that they are seen" (*The Birth* 108). Between seeing and speaking there is a necessary gap in which the eye/I is equally engaged in a process of in- and exhaling bits of meaning. In my opinion, persons/patients exist on the threshold of (not) seeing and (not) being seen. Yang's images of Allan dying give the impression that his body levitates, jutting out into space—but unfortunately without much meaning.

On the other hand, the posters advertised for AIDS are simple, if not quite embarrassing and disrespectful given the gravity of this illness. They rarely touch on any aspects related to the illness itself, as they allude more to the immorality of homosexual acts. Gilman explains part of the rationale involved in the process of not presenting people dying of AIDS as follows:

> For representing the ill body as a dying body is not possible. Such a body would point to 'deviance from the norm' in the form of illness. And this association with homosexuality and addiction labeled as illness must be suppressed. [...] All these images are images not of educating, but of control. (162)

The poster chosen for illustration reads "LOVE AIDS PEOPLE," with AIDS used as a verb and not as a noun; nonetheless, the construction's subtlety is rather counterproductive. To a certain extent, this poster reminds me of Michelangelo Merisi Caravaggio's *The Incredulity of Saint Thomas* (1601-02). There, the Apostle touches the actual wound because he needs tactile proof to accept its existence. The act of touching, as well as the skin open by the wound, reveal the fact that "Skin lacks the depth, the interiority we want it to give us. [...] [t]he flesh we crave as confirmation of our forms cannot do anything but turn us forever out even as we burrow into the holes we find there" (Phelan 42). But the poster presented above brings into focus *verbally* (therefore propagandistically) how one's body might be destroyed because of AIDS. Furthermore, the symbol of the arrow is a recurrent motif in the art representing AIDS, especially in light of its religious association with the martyrdom of Saint Sebastian.

But if LOVE AIDS PEOPLE, and if homosexuals identify themselves with a martyr, then they might easily fall target to this twisted logic and think of themselves as victims. As Kramer notes, homosexuals are tragic people partly because they feel responsible for an illness that has been affecting both the homosexual and heterosexual communities: "The continuing existence of HIV is essential for the functioning of the totalitarianism under which gay people now live. It works like this: HIV allows 'them' to sell us as sick. And that kills off our usefulness, both in our minds –their thinking we are sick—and in the eyes of the world—everyone thinking we are sick" (65).

How and by whom are our sexual identities created? Does the presence of one specific anatomical organ delimit one person's sexual identity? We have been trained into believing that there are only two genders, male and female, partly because of our binary way of thinking. Needless to say, just as in one color there are degrees of its intensity and saturation, so there are in us verbal, behavioral, and sexual tendencies that could make us look and act more or less masculine or feminine. Still, the borderline between these two is very fragile. Does a man choose to be(come) homosexual? Homosexuals have always been a target since, allegedly, they are a menace to the institution of marriage, procreation, and to morality in general. According to Jennifer Terry,

> Earlier studies from the 1930s aimed at determining distinct somatic features of homosexuals for the most part failed to produce any such evidence. Most of them focused on the overall physical structure of bodies, measuring skeletal features, pelvic angles and things like muscle density and hair distribution. (144)

In addition, endocrinology has not been able to say with certainty why some people prefer to engage in homosexual rather than heterosexual acts. All these studies only suggest that a human being—mistakenly labeled "deviant"—is a curiosity for those who consider themselves (or believe themselves to be) "normal." Terrifyingly, our attention has shifted to homosexuals; their bodies are unjustly displayed to our avid curiosity.

The importance of the erotic in our lives may be useful because the photos (and the minitexts) presenting Allan seem insufficient to initiate a dialogue by themselves. Because the eroticized body is what dies, that is what is put at risk or could become powerless because of AIDS. The body that cannot touch and be touched anymore; the body that cannot control its needs and desires; and, ultimately, the body that is deprived of its pleasures and thus loses its erotic self. Therefore, AIDS is not only a way to redefine our erotic life, but also becomes a reason to question our hygiene practices. Grosz points out that "[e]rotic pleasures are evanescent, they are forgotten almost as they occur" (*Space* 195). But when erotic pleasures are controlled, as seems to be the case because of AIDS, have we intervened in such a manner as to program our intercourse? Admittedly, AIDS is predominantly linked with one's sexuality and, hence, it could make one feel too self-aware about one's needs, as well as rigid and self-conscious in an (intimate) act which, in essence, is all about losing oneself, being uninhibited.

Also, let us not forget that the posters presented by Gilman did not use sex as sex, as a product; instead, "These images are images of holding and touching, and commodify the closeness of 'love' to justify the use of condoms. […] 'love,' here represented by the unclothed body, is the eroticized body" (Gilman 144). If the images presenting Allan were considered a product, and were accessible for the lay community, what would they advertise/sell? Are these photos (and similar others) an attack on (healthy people's) intimacy? Allan is never presented undressed. But his fragile, scrawny body is visible underneath his garments and/or the hospital bed sheets. However, what is not visible is Allan's skin, Allan--the person, Allan--the lover, and Allan--the friend. The repetition of his name may be disturbing, but it helps me point out even more emphatically how all these enumerated attributes are irretrievably gone. The photos cannot allude to a former Allan. Unlike what Siebers suggested (at the beginning of this section), namely that artists re-presenting AIDS are in constant role-variation--from subject to object and from object to subject--Allan is not subjected to this fluid role anymore and becomes exclusively an object in the intruded-upon privacy of his dying. In the end, Allan's sense of identity seems to be imprinted only in the camera's objective lens. After he died, as Yang remembers, "I read his diaries […]. AIDS was a tragedy that was for sure, but as well he had an addictive personality and his day to day life was full of desperation. I hadn't realize the extent of this and it came as a shock. Yet there were moments of clarity when his fresh test for life shone" (51). Yang does not say more about Allan's intimate writings and, as he suggests, it was quite surprising for him to discover a richer, more intimate dimension of his friend. Still, until Allan's diary/diaries will be released to the public and, by being available, we might have a more palpable view on his life, we rely exclusively on the selections of photos and minitexts accomplished by Yang, thus being aware that, no matter how exquisite they are, they could say only a few things about this enigmatic person/patient.

Jo Spence

In the first two sections of this chapter I introduced certain works of Wojnarowicz and Yang as they focused on AIDS as a complex of illnesses. According to Sontag, "AIDS—acquired immune deficiency syndrome—is not the name of an illness at all. It is the name of a medical condition, whose consequences are a spectrum of illnesses" (*AIDS* 16). Cancer is another controversial word. There is not one single definition for it because there are several types of cancer with different topological manifestations. To this confusion should also be added the fact that one is diagnosed as having cancer at a particular stage of the illness. In this section I focus on some artworks by British photographer Jo Spence, recording visually her reactions to cancer. My main concern *vis-à-vis* these photos is to see if they follow *or* defy Jacques Lacan's idea according to which "[t]he function of the image screen is [...] to filter the real: it protects the subject from the gaze of the world, [...] and tames it *in* images *as* images. Lacan terms this primordial function of the image screen a dompte-regard, or 'taming of the gaze.' [...] [t]o tame the gaze is not to block it entirely: it is to deflect it, to redirect it, as a mask does" (Foster 279-80).

When diagnosed with breast cancer, Spence did not opt for mastectomy; instead, she chose a treatment combining traditional Chinese medicine with phototherapy. In her works, Spence tried to juxtapose words and images, or, as she writes, "[w]e need to use our cameras, tape recorders, diaries, poems, videos—whatever cultural resources we have—to witness our own histories, to learn to protest and share, and to learn to nurture ourselves" (*Cultural Sniping* 140). Based on her philosophy that domestic photography, staged as it is, conceals rather than reveals our most intimate fears, Spence used her camera as a visual diary, accompanying her through the ordeal of cancer. As she remembers, "When I learned that I had breast cancer, [...] I used my camera as a third eye, almost as a separate part of me which was ever watchful: analytical and critical, yet remaining attached to the emotional and frightening experiences I was undergoing" (*Cultural Sniping* 130).

Could a body be placed in front of a camera in order to faithfully record its experience of cancer? When Spence learned about her breast cancer, she did not know how to make it visible and meaningful to herself and others. Her task was probably even more challenging, since she had to question her philosophy vis-à-vis her own profession. As she writes, "Family photographs hide any evidence of illness or ageing, since photographic conventions encourage us to 'smile for the camera' and the lack of clarity in small images prevents us seeing fine detail" (*Putting Myself* 155). How many of us could identify ourselves with the photographs taken in a studio, at a party, social event, or banal gathering? These photographs are not accurate for they are staged, controlled, and fixed. We manipulate the very instant of our visual recording in order to enhance our appearance. In his book, *Image, Music, Text* (1977), Roland Barthes affirms,

> The type of consciousness the photograph involves is indeed truly unprecedented, since it establishes not a consciousness of the *being-there* of the thing but an awareness of its *having-been-there*; [...] its reality that of the *having-been-there*, for in every photograph there is the always stupefying evidence of *this is how it was*, giving us [...] a reality from which we are sheltered. (44)

Nonetheless, while a photograph may testify to our "being there," the time represented counterfeits our seeing, and, even more significantly, our memory of the moment. Therefore, understanding this clear distinction between photographs as staging an identity and photographs as reflecting an identity, Spence realized that her body could never be something other than a transitory site, onto which its fleeting signs alternate their attributes constantly: inside/outside, personal/social, and meaningful/meaningless.

Furthermore, when diagnosed with cancer, although fully aware that neither words nor photographs (or any other media for that matter) could accurately testify for the body in pain, Spence nevertheless exposed her body in its partial or full nudity, thus sculpting out of her body many slices of fear and suffering. Entitling her project *Narratives of Dis-ease: Ritualized Procedures* (1989), Spence voices its manifesto:

> In these photographs is the beginning of a 'subject' language, one which allows me to start the painful process of expressing my own feelings and perceptions, of challenging the 'ugliness' of being seen as Other. In so doing, I cease to be a victim, becoming again an active participant in life

> [...] If I don't find a language to express and share my subjectivity, I am in danger of forgetting what I already know. (*Cultural Sniping* 135)

The human body of any ill person is a powerful, evocative site and case in point for all the transformations a body experiences. Spence's body acquires a reading of monstrosity: what used to be healthy is now transformed into the unrecognizable Other. The question that Spence might ask is: What has happened to my body? Have I become overnight (here time is instantaneously contracted!) a stranger captive in an altered narrative and bodily image? Because of her deformed left breast, as well as of the half mask she wears on her face, in this photo Spence speaks about her real "monstrosity." It is not her sentient body that has become monstrous; it is, rather, society's obsession with how the body should be/look and its refusal to acknowledge her suffering. In a desperate attempt to be seen, onto this woman's body is written "monster" (an explosive example for the dichotomy between visuality and textuality, as well as the body as seen, read, perceived and interpreted). Moreover, although not only the act of speaking about a terminal illness but also the act of looking at one's deformed body is considered taboo, photographs finally permit us to analyze and construct different discourses about the body in pain.

For example, in another photo, Spence inscribes on her cancer-affected breast "Property of Jo Spence!" What could be the meaning of this very unusual, yet powerful inscription? On the one hand, Spence testifies that her body, despite having been recently diagnosed with cancer, is still her own, her belonging, her property. But on the other hand, this image could also function as archival memory for the body that could lose a breast to mastectomy, or could even die because of cancer. As Spence recounts, "Before I went into hospital in 1982 I decided I wanted a talisman to remind myself that I had some rights over my own body. This is the one I took with me. I felt I was entering unknown territory and wanted to create my own magic fetish to take with me" (*Cultural Sniping* 120). Thus, by exposing her body, Spence wanted not only to make it visible to others, but also to overcome those trends according to which it is shameful to show publicly the transformations of the body in pain. As she writes, "A lot of my work is about overcoming shame, about speaking what people feel has no validity, no right to be heard. [...] Shame is a cultural construct: one is not born ashamed, one is made so" (*Cultural Sniping* 213). Throughout her works Spence asserts over and over again that just as there are sanitized photographs so there are sanitized discourses that can profoundly harm a person, and, by extension, a society. By her act of refusal to have a mastectomy, Spence undermines not only the socio-political discourses of women with breast cancer, but she also shows that medicine is not ready to embrace and/or combine other methods of treatment. As she sadly admits, "I do not think I have been so lonely in my entire life as I was after I'd refused traditional allopathic treatment for breast cancer—the mastectomy and radiotherapy" (*Putting Myself* 198).

The act of speaking is/should be a moment of enchantment; understood as drugs, words are healing or harming, revealing or camouflaging, constantly taking us by surprise as long as they invite us to participate in a dialogue. While we speak with a certain choice of words and tone, medicine is still in its incipient, prelinguistic stage. Medicine does not treat us as persons but as "cases," not individually but collectively. The discourse of medicine is schematized, fleshless. Although a word may have a certain impact on one person, but none on another, medicine still employs the same discourses over and over again. Thus, its discourses are out of context, misplaced, and impersonal, and its signifiers are disembodied, floating aimlessly.

Consequently, pain insulates us. If we are more or less concerned about our bodies when they are healthy, we are definitely shattered when the moment of becoming ill furtively approaches us; when voicing pain from inside out, our bodies roar with anger. Moreover, the drama of an ill person/patient has to be staged outside, inscribed onto the volatile surface of one's body, in order to *contaminate* and catch the others' perception/attention. We are an amphitheater of malleable embodiments, polyphonically arranged, constantly drawing the "blueprint" of our bodies, not knowing if our bodies will obey or disobey our needs. But the moment of becoming ill should not correspond with the moment of remaining silent. Becoming a patient is a rhetoric of a body about or attempting to be known. As Spence writes,

> In order to understand we first of all have to feel safe enough to deconstruct. We then have to put the pieces together again and again, continually montaging until we make new connections which will enable us to break out of the psychic bonds which hold us. Out of the broken pieces of the self will come a subjectivity that acknowledges the fragmentation process, but which encompasses and embraces the parts and brings them into dialogue with each other. (*Cultural Sniping* 122)

In an attempt to express her thoughts via photographic means, Spence created one of her most fascinating works. The image reads, "How do I begin to take responsibility for my body?" Sitting naked in front of a mirror and using paint on some parts of her body, this image is just as shocking as it is rhetorically powerful. In front of a mirror, a body does not see itself. Put differently, a mirror does not project a reflection of our image as a whole, but rather it reveals the body as a hole which, when pierced into, leaves behind the echo of its multifarious voices and desperate cries. In addition, the image that is temporarily reflected on a mirror functions like a two-dimensional painting, thus lacking depth. Suffice to say, our bodies are *tempores*, a semantic hybrid, a portmanteau, a jocund combination of temporal and pores, just as time transpires on the surface of our skin, permanently connoting our passing through life.

We live underneath our somatic, affective, and cognitive plateaus, or as Deleuze and Guattari explain this process:

> Gregory Bateson uses the term *plateau* for continuous regions of intensity constituted in such a way that they do not allow themselves to be interrupted by any external termination, any more than they allow themselves to build toward a climax. [...] A plateau is a piece of immanence. Every BwO is made up of plateaus. Every BwO is itself a plateau in communication with other plateaus on the plane of consistency. The BwO is a component of passage. (158)

Like an archeologist who digs deeper and deeper into the soil so that s/he may restore the remaining parts of an artifact, philosophers such as Deleuze and Guattari try to deconstruct the human body. Furthermore, after restoring the pieces of an artifact, an archeologist glues them together, so that, through them, he might restore a part of a population's (lost) civilization. But what would be Deleuze and Guattari's project? Briefly put, the two French philosophers emphasize that our organism is situated at a crossroads of feelings and perceptions, and, consequently, "[d]ismantling the organism has never meant killing yourself, but rather opening the body to connections that presuppose an entire assemblage, circuits, conjunctions, levels and thresholds, passages and distributions of intensities" (161).

> But where could we possibly locate the distribution of intensity in Spence's "How may I begin to take responsibility for my body?" As she writes, How do we deal with the abject loneliness of the long struggle for health (the most boring of subjects to other people who are 'well')? How to present yourself as a subject in daily struggle? People are used to the 'narrative resolution' of illness like cancer (in the media's terms you are either 'dead' or 'better'). (*Cultural Sniping* 130)

What Spence may assert is not only the suffering generated by having a terminal illness, but also the solitude of the ill person/patient and the fear of communicating his/her pain to others. Moreover, when she exhibited the images of her body in pain, there were no reactions from the audience, as if those in the audience were struck by an unmotivated silence. How could we constructively problematize the body in pain if we reject it? As Spence records,

> I looked for a discourse of cancer in the newspaper: I wanted to see how the medical profession sold its truths. […] very quickly [I] could discern that what passes for 'news' is in fact a series of public relations exercises to keep certain ideas in the public domain which […] help reinforce the power of the medical and pharmaceutical industries, but at the same time they also help to whip up more and more hysteria about the disease. (*Putting Myself* 188)

She goes on to say that our notion of health (like all other notions) comes via mediated sources, and as a consequence, despite years and years of accumulated knowledge, we are still patients who are acted upon rather than acting out our most fundamental needs and rights. Therefore, I believe Spence's legacy is an attempt to open up these secularized and ineffectual discourses that damage our bodies and minds in very subversive ways. Spence points out that photographs "[a]ct as a tangible marker of something which could otherwise go back into the unconscious and remain dormant for a long time." (*Cultural Sniping* 96). Put in another lexicon, when archived, photographs could offer us a more profound way to understand our bodies in motion, along with their several embodiments. We may feel the desire to hold these photos in our hands, and, in return, they hold us back into a past with more or less accessible paths. Spence's photographs proved that this art needs to be supplemented by a powerful, personal narrative in order for their stories to transcend their initial stage of development in the *camera obscura*.

Epilog

In this chapter we have seen people exposing their nude body, or the dying/dead body of their beloved. What is the impact of their exposed skin on us? Connor believes that "The skin marks time partly by effacement: by the healing of lesions and reassertion of the surface against every assault. […] No other feature of our physical lives offers so magical a promise of reversibility" (46). The photos included in this chapter present the moment of the irreversibility of our skin when it loses its "magical" capacity for regeneration. Nonetheless, the artists insisted on shooting this fleeting moment of their (or their beloved's) embodiment because they wanted to show how they were still alive.

Then, if it is problematic to redress a dys-appeared body, for a body does not have a standard functioning but fluctuates despite our efforts, what Wojnarowicz, Yang, and Spence attempt to establish is a normalization and/or eradication of the socially dys-appeared body. Their campaign at showing the ill, dying, or dead body contributes to or reflects a current cultural trend: our refusal to accept that only physicians are in charge of curing an illness. As Sontag remarks, "[o]nce, it was the physician who waged *bellum contra morbum*, the war against disease, now it's the whole society. […] [t]he war against diseases […] are more money to be spent on research. […] Victims suggest innocence. […] And innocence suggests guilt" (*AIDS* 10-11). Pursuing this line of reasoning, we return to the "scapegoat motif" encountered earlier when we discussed some of the rhetorical techniques employed by the Pro-Gay Movement. However, Sontag does not name the victims; consequently, we could list in the category of "innocent victims" not only those who die because of AIDS, but also those who have cancer, Alzheimer's, Parkinson's, etc. A study about funding research on these illnesses would generate sensitive political tensions.

However, with this idea in mind, Sontag's "bellum contra morbum" reminded me of Kushner's emblematic character, Roy Cohn, from *Angels in America* (1994). Cohn is the kind of person who re-presents an illness from its political angle. He suggests that as long as one has power, one can more easily interfere within the medical infrastructure, obtain better treatments, and/or other facilitated services.

Yet when one is caught in continuous and agonizing pain, what matters the most? Is it the comfort brought by friends? Is it the soothing effect of a tranquilizer? Is it the thought of an afterlife? Or is it the plain truth of our dying? Roy chooses mentally the last "option," while physically being helped by some very efficient tranquilizers. Once again AIDS has struck not a specific person, but a nameless body. The only thing left for Roy to do is to control pain through his connections:

> ROY. Oh hi Martin. Yeah I know what time it is, I couldn't sleep, *I'm busy dying*. Listen Martin, this drug they got me on, azido-methalato-molamoca-what-chamacallit. Yeah. AZT. *I want my own private stash*, Martin. (1.6.31, emphasis added)

In the light of Cohn's understanding of death, as well as of the three artists' works presented here, is dying/death something that reveals our self-centricity? It is by now a truism to say that death is the final moment of our embodiment to which we are denied access. Nonetheless, we cannot stop thinking about (our) death, and the last passage of this chapter proposes its own reflection on this subject. Elias argues that each one of us is a *homo clausus* (Latin for "closed, self-sufficient being"). He believes that this condition is a consequence of our living an advanced phase in our individualized life. Surprisingly, he relates this self-sufficiency to the ritual of dying. He believes that in highly industrialized societies, a person/patient may benefit from the most recent technical and medical equipment, but that that person usually dies alone, meaning without his family/relatives around him/her. On the other hand, as he goes on to argue, "[f]amilies in less developed states [...] often go hand in hand with far greater inequalities of power between men and women ... [The dying] take leave of the world publicly, within a circle of people most of whom have strong emotive value for them, and for whom they themselves have a such a value. They die unhygienically, but not alone" (87). Elias does not explore this idea in depth, so we are left to wonder what he meant by dying unhygienically, or if he thought that method was better in coping with death. Also, he never mentioned the exact countries/regions he had in mind when he made that remark; therefore, we are left unsatisfied by his comment. Nonetheless, as Elias reminds us, it is important to remember that the traditional death rituals were and are intimate moments (and they should remain like this). The homo clausus idea may be linked with a body that is reaching its final embodiment, and hence becoming a closing-in-itself body.

However, how does a body transact and/or negotiate the moments of its final embodiment? The process of sinking and/or dys-appearing in one's body, to which I refer, is not a visually, aurally, or especially olfactorily pleasant experience. I have come to this conclusion based on the theories presented by Mary Douglas' study, *Purity and Danger: An Analysis of Concepts of Pollution and Taboo* (1966), where she demonstrated that the body leaks at its margins. For Douglas, our bodies are anything but closed systems. One may, however, suggest that our bodies are corpora aperta within which—it seems—only our deceitful memory misdirects our emotional brains by indicating which subsystem is still functional and open and which has become useless, that is, closed. In this light, we should redefine Elias's idea by saying that what appears to be a monolithic structure, i.e., a "body-closed," sealed, and/or self-contained, is in fact a very fluid body; that death does not reveal our self-centricity because that reasoning may generate an absurd idea, namely, we die alone because we have spent a life alone. Consequently, I argue that the dying body becomes the margin par excellence, which, because it is completely out of control, does not stop from leaking and/or emitting smells. According to Lawton, "[o]n a number of occasions, staff kept aromatherapy oil burners running throughout the day and night in an attempt to veil the odour of excretia, vomit and rotting flesh. [...] I observed that smell created a boundary around a patient, repelling others away" (135). One has to close one's eyes to dimly/vaguely imagine what it must feel like for the medical personnel to keep the vigil of the dying bodies.

Yet, how come the lay community is exposed to photographs of the dying only on rare occasions? Why are they kept hidden? According to Gilman, these images are not made public because "The classical model of 'healthy/beauty' and 'illness/ugliness' is part of a cultural baggage that accompanies any representation of the ill or healthy body." (118-19). In the absence of the healthy body, most likely the photos presented in this chapter could not help their subjects relapse into their memory, where they would have traced backwards the syncopated path marks of their spent lives. While the skin is endowed with the capacity of regenerating itself after it has been wounded, thus effacing time, a photograph of a dying body seems to efface one's memory of one's accumulated experiences. Such a photograph makes its contents (that is, the time, location, personal context of the shooting) disappear since its details will eventually fade away. As a corollary, the absent body effaces its photographed version, leaving it few chances to be remembered. The theme of the *ars moriendi*, as presented in this chapter, has demonstrated that what dies is not only one's body, but also the echoed memory of its erotic self.

CHAPTER SIX

MOLECULAR PROXIMITY

> You are more and more authentic the more you like someone you dreamed of being.
> (Almodóvar)

I started this book with chapter *Le corps perdu/Le corps continué*, where I stressed the necessity of creating ways for lasting (either through artificial life, artifacts or prolonging one's life through organ transplant). The last chapter of this study suggests that, in trying to define the current condition of the human body, there is yet one more aspect that needs to be added. More specifically, one of the bold questions addressed in Sontag's book *Illness as Metaphor* is whether or not a society could catch a (fatal) disease. While Sontag is particularly focused on a society in which its citizens are diagnosed with cancer, this chapter takes a look at individuals whose bodies lose their human carnality and become animals as in Franz Kafka's 1915 *The Metamorphosis* or return to their human shape after an animal phase as in Moacyr Scliar's 1983 *The Centaur in the Garden*, and how this metamorphic process affects a society.

The main argument that connects these novels is based on the notion of the "dys-appearing body." As defined by Drew Leder in his book *The Absent Body*, in such a state "[t]he body *appears* as thematic focus, but precisely as in a *dys* state—*dys* is from the Greek prefix signifying "bad," "hard," or "ill," and is found in English words such as "dysfunctional" (84). Leder argues further that a body could "dys-appear" (as opposed to "disappear," for example, the body at partial rest, during sleep) when one is diagnosed with a terminal illness as well as when one finds oneself in severe hunger, thirst, and/or weakness. How could we better define this phenomenon? What are its implications? How does one exist in a body that is dys-appearing? These are some preliminary questions that frame this chapter.

It is worth noting that just as there occurs a physical dys-appearance of the body in severe pain, there might also be a social dys-appearance. In the second type of malfunctioning, "[t]he split is effected by the incorporated gaze of the Other. […] A radical split is introduced between the body I live out and my object-body, now defined and delimited by a foreign gaze" (Leder 96). Those in abominable pain are confronted not only with their acceptance of having a body different than it used to be, but also being unjustly marginalized, isolated in hospitals, asylums, et cetera. What Leder may imply is that the gaze of the other thematizes the body. To take the issue of thematization a step further, we could suggest that illness itself thematizes a body. We do not (or rarely) see our own body because we are sculpted into it, and consequently barely notice its evolution (or should I say involution?). But how could we exist outside of our body and still somehow be aware of it? Does only the encounter with an illness remind us that we have a body?

Furthermore, whether in good or poor health, how do we perceive and transform our body into bodily narratives? "In an age of volatile extremes," such as ours, "the body has become another realm of fantasy" (Anker 185). We are assembled and assembling machines, made of congruous as well as incongruous parts. Memory, and along with it the sum of our fulfilled and unfulfilled needs, plays a substantial yet tricky role in the assembling of life. We remember events distortedly because time melts away the fringes and, more often than not, the contents of our narratives. The closer we attempt to move toward the epicenter of our lives, the more difficult it becomes for us to untangle its labyrinthine structure. For Merleau-Ponty writes, "My body is the fabric into which all objects are woven, and it is, at least in relation to the perceived world, the general instrument of my 'comprehension" (273). But what happens when the fabric of the body breaks? As we shall see, the literary works chosen for interpretation focus on two examples of a "body-broken"; they also form a pertinent example by adding another component to suffering and dys-appearance, i.e., mutation. In these works by Kafka and Scliar, pain is visualized either as a metaphorical exhumation of the animal lurking beneath the well-camouflaged site of the human or the human reemerging out of the animal cocoon.

To understand this transformation, or breakage, it is worth noting this question: How do we define pain? According to Greenfield, "[w]e know pain is expressed as other associations: pricking, stabbing, burning, chilling. […] We [also] know that the more people anticipate pain, the more they perceive it as painful; and I would suggest that that is because there is a build-up of neuron connections" (19). We may say pain is one constant in our fleshy ontology. When in pain, the body--or the affected bodily area--is thoroughly sensed by us, giving us the impression we are locked in our body. This phenomenon is known as "coenesthesia," or "[t]he general sense of bodily existence. [However, when in pain,] we are no longer dispersed out *there* in the world, but suddenly congeal right *here*" (Leder 75). Particularly related to the works discussed here, regardless of how these authors decide to approach the theme of pain *as* bodily mutation and imprisonment in "here," the animal (centaur) and insect (bug) involved in these transformations are terrifying, disturbing, yet eerily fascinating. The outcome of combining and grafting human tissues onto their animal counterpart brings into focus the hybridity of the project, along with its ethical questions.

For the moment, suffice it to say that the theme of mutation has ancient origins. In Greek mythology, a chimera (and chimera-like creatures) was considered a hybrid animal, having the body of a goat, the head of a lion, and the tail of a serpent. As Suzanne Anker and Dorothy Nelkin write, "[c]himeras once populated the literature, mythology, and art of ancient Greece. They were dangerous, formidable, and powerful beasts, representing fantastic yet uncivilized and chaotic forces in nature that confronted mankind" (107). However, to ground the discussion of this theme only in literature, would be to limit its vast implications. Instead, in analyzing the experience of pain of Kafka and Scliar's characters, we may learn not only how to be more sympathetic toward the pain of the other, but also acknowledge that "[t]hrough sound, through the various refrains, we invent, repeat, and catch from non-humans, we receive news of the cosmic energies to which we humans are always in close, *molecular* proximity" (Bennett 168, emphasis added).

We may have actually crossed the line of "molecular proximity" and redirected ourselves toward a borderline physical identity. Transgenic experiments, for instance, have been conducted since the end of the World War II. Anker argues how "Bioengineers have created sheep-goat chimeras, known colloquially as "geeps," and transgenic pigs that produce low-cholesterol meat" (90). Ironically, while our carnality has been deeply influenced and/or altered by scientific experiments that keep it at a debatable ethical intersection between what is human and what is animal, when in pain, we have not yet learned how to howl at the moon. In other words, to regulate our pain, we depend almost exclusively on medical prescriptions; yet, as I show here, there is a social component of pain that cannot be treated with drugs.

Gregor Samsa Had a *Dream*

We come into this world on a cradle of dreams, and we continue to dream throughout our lives. We keep our eyelids shut for a while, hoping to reach beyond the limitedness of our given spatial-temporality. We feel a thrill at the mind's fracturing and relocating itself toward places saturated with our creating and populating of the world. A few dream to destroy and harm ourselves and others, but even fewer awake to find their human, vibrant carnality transformed mockingly into the disturbing shape of a gross insect. This is the experience Kafka's main character in *The Metamorphosis* undergoes: "As Gregor Samsa awoke one morning from uneasy dreams he found himself transformed in his bed into a gigantic insect. […] What has happened to me? he thought. It was no dream (89).

Shocked by this transformation, he desperately clings to his mind, the last vestige of a once used-to-be-exclusively-human site: "And at all costs he [Samsa] must not lose consciousness now, precisely now" (93). Does the human, recording yet self-effacing mind, reel backward, hopelessly trying to unearth the meaning of the human condition? What type of dys-appearance does Samsa experience? Kafka does not offer much about the origin of Samsa's transformation. Nonetheless, his character's experience is profoundly shocking; one day he wakes up differently, not as a human being, but in the shape of a bug. What Kafka may imply is that his character suddenly realizes he cannot use his body any longer, and thus feels it as a destitute, yet burdensome-to-wear *shell*. As Merleau-Ponty suggests, in such unfortunate cases,

> [i]f the world is dislocated, this is because one's own body has ceased to be a knowing body, and has ceased to draw to get all objects in its own grip; and this debasement of the body into an organism must itself be attributed to the collapse of time, which no longer rises toward a future but falls back on itself. (329)

Furthermore, as he sadly acknowledges, Samsa suffers a double shock, if not an insurmountable disappointment. On the one hand, he is confronted with a new carnality whose nature not only alarms him, but also of which he does not know much, let alone accept. On the other hand, he must face the disgust and shame of his family when they see him:

> Slight attacks of breathlessness afflicted him and his eyes were starting a little out of his head as he watched his unsuspecting sister sweeping together with a broom not only the remains of what he had eaten but even the things he had not touched, as if these were now of no use to anyone. (Kafka 108)

If Samsa suffers from an enigmatic mental and corporeal collapse, and taking into account his family's cold and unsupportive reaction, could we go further by suggesting that anomalies function similarly to a contagious disease? Or why do we believe that the body without anomalies would remain unchanged? The body is in continuous motion and, therefore, transformation.

Unfortunately, whether at night or during the day, Samsa does not experience the space surrounding him as being infinite. On the contrary, he may feel this space and his presence in it as closing more and more with each day passing. Thus, not comfortable with his new condition and without support from his family, Samsa thinks he does not have any option other than to accept his suddenly undesired transformation. As a consequence, he maddens himself: "[h]e began now to crawl to and fro, over everything, walls, furniture, and ceiling, and finally in his despair, when the whole room seemed to be reeling around him, fell down onto the middle of the big table" (119). He repeats this agonizing ritual until deciding to put an end to his already physically and socially dys-appeared body by committing suicide:

> The decision that he *must disappear* was one that he held to even more strongly than his sister, if that were possible. In this state of vacant and peaceful meditation he remained *until the tower clock struck three in the morning*. [...] Then his head sank to the floor of its own accord and from his nostrils came the last faint flincher of his breath. (135, emphasis added)

Samsa suffers from an inexplicable mutation whose etiology baffles everyone. He does not have a human body anymore, but rather bodily *sequelae* which, having been pressured by physical pain and social alienation, have resulted in his current deteriorated state of being.

However, Samsa seems to be the medical enigma *par excellence*, or the "non-case" of medical cases. Could Kafka, then, be arguing that treating an illness implies an unavoidable degree of isolation, a distortion of personality and a mutation of identity? Since we do not know the exact nature of Samsa's transformation, does his situation imply that we could be transformed against our will? One day, could we become Samsa, that is to say, unrecognizable to ourselves and others? If pain validates a passing through life, and if the encounter with pain cannot be minimized, what should be the next step? According to David B. Morris, "[t]he behaviorist begins from the proposition that all pain can be redefined as pain behavior" (142). For a better understanding of pain, he suggests a suturing between physical and mental pain. In doing so,

> We [would] recover a sense of the importance of minds and cultures in the construction of pain, and we [would] begin to proliferate the meanings of pain in order that we do not reduce human suffering to the dimensions of a

> mere physical problem for which, if we could only find the right pill, there is always a medical solution. (Morris 290)

However, for Samsa there is no available drug to alleviate his agony. Perhaps unintentionally, Kafka offers a further step in the concept of dys-appearance. His character commits suicide, thus totally disappearing. In this light, it is relevant to note the differences between the two concepts: while dys-appearance implies a body still somehow alive, disappearance includes death.

The Centaur

While in Kafka's story, Samsa simply, though inexplicably, awakes in the shape of a bug, in Scliar's novel *The Centaur in the Garden* there is a more intricate scenario. By the time we complete the reading, we discover that there is no clearly defined identity for the main character. What this author proposes for himself as a writer--and for us as readers--is to imagine a character whose inner and outer contours, dimensions and perceptions are never fixed, but mutable and interchangeable. Moreover, he also suggests that when realities are insufficient to understand who we are, we create myths, novels, and the like. As Scliar suggests, we are more than humans; we are centaurs, Sphinxes, or any other mythological creature superimposed imaginatively on our human identity. In this scenario, life seems unfathomable, unreachable, and unpredictable: "We must contrive to understand how, at a stroke, existence projects round itself worlds which hide objectively from me" (13).

When Guedali, the main character, was born, he had an uncertain biological identity, half-human, half-animal. Shocked and disgusted, his mother could barely look at him, let alone nurture him: "Her [Guedali's mother] silence is an accusation to her husband: It is your fault, Leon. You brought me to this place at the ends of the earth, this place where there are no people, only animals" (65). In order to understand the mother's bitter reaction, it is relevant to note that she and her husband were Russian immigrants in Brazil. This raises the question of whether the place of our birth is pure coincidence, something accidental? Or to the contrary, does it engrave within us the social "genes" of our cultural identity? All those who decide to leave their country of birth experience various degrees of nostalgia, metaphorically translated into mutations and emotional mutilations of various kinds.

For Guedali, the problem is yet of an additional nature. Having been born hybrid, he is equally confused and fascinated by his physicality. He finds relaxation, ease, and pleasure in galloping for hours over the vast Brazilian pampas: "This gallop. This gallop in the middle of the night, through open fields, through swamps that reflect a pallid moon, this gallop is to stay in my memory for many years" (108-109). When he does not gallop, he reads feverishly, searching for an ancestor to his blessed or cursed fate, but his endeavor is futile. He does not find anything or anyone even remotely related to his unusual condition. The more he searches, the more frustrated he becomes. Finally, he embarks on a cruise to Morocco, where the lower part of his body will undergo a dramatic surgery. The surgery proves to be a success, if not a medical marvel; the centaur's marks of his identity are surgically removed and replaced with human equivalents, for example, hooves with human feet. Still, he has to wear trousers and boots to protect areas of his body not yet fully recovered. For this reason, he feels that someday someone will find out the truth: "What is, one day, I had to have an urgent operation? [...] Or if someone spied on us inside the house, with binoculars? (As to that, I had already taken the precaution of buying thick curtains)" (175). What Scliar may be implying is that, despite the sum of our efforts to fit into society, we never know for sure when and/or how to please it. After surgery, and much to his disappointment, Guedali discovers that people who have been born without an anomaly, but are poor and disfranchised, are considered freaks too—a category from which he thought he had escaped surgically.

Baffled by an unwelcome reality which he could not fully anticipate, Guedali does not feel completely fulfilled or protected at home either, an experience he "shares" with Kafka's Samsa. He is deeply confused. Are his surgery and sacrifices validated? Instead of finding soothing answers to his problematic ontological questions, he becomes more and more alarmed and agitated. He hopes that by reviewing his past life, he will find the link missing from his "evolutionary" process, and thus be able to attach it and recover his identity. He discovers the opposite:

> A centaur, with great effort, becomes a man; later, he goes through a crisis and decides to become a centaur again; then he does not really know if he wants to be a centaur again or not; meanwhile, he meets a crazy sphinx who wants to transform him into a lion man! Ridiculous. Mythological delirium. (7)

What he uncovers is that at the foundation of a person—as well as a society—there are dreams, but there are nightmares, too. There are actualized, mature desires, but there are also repressed desires that (could) return as agonizing pains. The mythological delirium is just another way of pointing to the state of confusion that permeates the fabric of our body as well as society.

In the end, does Guedali choose to remain human or centaur? His epiphany reveals his nature. Guedali is not just a combination of animal and human, but a symbiosis of elements, manifested in one: vegetal, mineral, animal, and human—just as we are since these four elements have altered our shapes and personalities. In this context, an emblematic passage that defines Guedali occurs when he encounters Lolah, the Sphinx. He declines to answer her famous ontological question; instead he releases her in ecstasy by pleasuring her. Above all, Guedali is a man of the senses. Indeed, his inner world is sensual.

However, if we think that sensuality is the tone with which Scliar concludes the novel, we are bitterly mistaken. The last chapter functions as a contra-argument to the narrative developed thus far, or perhaps it is metafictional—combining and engulfing a story within a story. Consequently, the overall structure *vis-à-vis* the book is like a vortex, centripetal, a *mise en abîme* provoked by Scliar with one possible intention: to let us discover the multifold narrative dimensions of a literary work.

Just as our coming in and out of life is a mystery, so it is the process of reading and making sense of a book: "One's first memories naturally cannot be described in conventional words. They are visceral, archaic. Larvae in the heart of the fruit, worms, wriggling in the mud. *Remote sensations, vague pains*" (Scliar 175). Memories and realities mix together, fictionalizing themselves. Anything could potentially happen, even in a novel. Scliar twists the end of his novel so drastically that, when one closes the book, one is left dumbfounded: Was Guedali a man? A centaur? Or perhaps a cancer patient? In the last chapter, Tita--Guedali's wife--narrates the story from her perspective. Again, as in Kafka's story, the style is ambiguous. Apparently, Guedali's "Morocco story" took place when he had surgery related to cancer, though we are not given details about it. To deal with chronic pain, and most likely the effects of medication, Guedali experiences amnesia by projecting himself as a centaur. As Greenfield points out,

> Pain is absent in dreams, and that is a small assembly state. Similarly, morphine gives you a dream-like euphoria: the way morphine works is via a natural opiate which acts to make the brain cell assemblies less efficient at being corralled up—therefore, people will often say that they feel pain but it no longer 'matters,' it is no longer significant to them. (19)

Even bodies in severe pain and physical deterioration could (and some actually do) continue to function because they have mastered the language

of habit, even though they now perform it with their last physical resources. The well-known chemist Ilia Prigogine spent a good portion of his research focusing on what he called "far-from-equilibrium" systems; for these unstable systems "a fluctuation can become a bifurcation point or a site of a swerve, where the system spontaneously chooses a path that no scientist can predict" (Bennett 101). Furthermore, "Prigogine contends that, even in its most complex and indeterminate state, a physical system continues to possess a kind of intelligibility."
(Bennett 103-104). If bodies in severe deterioration could be read as "far-from-equilibrium" systems, then this kind of intelligibility is a consequence of them having been engaged in performing habit.

However, Guedali might have wanted to forget about his illness. "Nociceptive" refers to something that causes pain. Or according to an online cancer web dictionary, it is "[t]he process of pain transmission, usually relating to a receptive neuron for painful sensations."[1] In taking these definitions into account, Guedali might have deliberately re-created his new identity as a centaur because his experience of cancer had taught him new habits. After all, surviving cancer (or any other terrible misfortune) implies accepting being different, that is to say, changed. In Guedali's words,

The story [Tita's] is as ingeniously woven as a soap opera. With one single objective:

> to *convince* me that I never was a centaur. [...] I still see myself as a centaur, but a centaur growing constantly smaller, a miniature centaur, a microcentaur. [...] Maybe it would be better to let him [the centaur] go, to accept this reality they want to impose on me: that I am a human being. (Scliar 214)

For most of the novel's readers, the ending is what disappoints them: why would an author want to recreate a totally different story at the end? For many readers, the ending of the book is an anticlimax. For other readers (as I), the ending, with its story turned upside down, is the beginning of traveling beyond what is in the book. Still, no matter which interpretation prevails, doubtless we can agree that by the end of Scliar's book, our hearts are racing faster, just as fast as Guedali's hooves galloping across the plains, running away from the "normal" people who are finally *just* human. Ironically, the idea of "normality" proposed by Scliar is impoverished; those who have forgotten to be mineral, vegetal, animal, human-- and I would add textual--are limited, the "abridged normal." Or, as Pierre Baldi writes in his book, *The Shattered Self: The End of Natural Evolution*,

> [i]n ancient times we thought that we were made of a stuff completely different from say, rocks, or that we were very different from other animals or plants, and that we occupied a special position in nature. [Today we know that] *homo sapiens* is made of atoms, like all other macroscopic objects we know. (11)

Can we ever escape our fear of living life as freely as possible? To answer this question, there is one myth that needs to be destroyed, and that is the myth of happiness. We have been told that this life is nothing more than a series of pains cut short by joys. At the end of life, there is the afterlife with its uncountable blessings. Why do we still follow this myth today? Happiness, like our passing through this life, is *not* a myth. *Heureusement*, Guedali chooses (and teaches us) otherwise. According to him, the flesh and blood of our ontology is the sum of what we have desired and dreamt about, achieved and failed. In his particular case, the moment he seizes his chthonic structure, he may begin to be happy and complete:

> I began to think about buying some land, if possible near the place where we had our farm [as a child]. [...] What I wanted was contact with the earth—an experience that I considered profound, visceral. I wanted to be barefoot, I wanted to grow calluses on the soles of my feet, to make them even tougher, even more like hooves. (214)

The dys-appearance of Scliar's Guedali, however, reads differently than that of Kafka's Samsa. After years of searching for validation for his mixed and complex identity, Guedali dys-appears in his essence by his own will. In this "place," dreams and nightmares about his identity may return, but he seems to have accepted them.

There is yet one more piece that needs be added to decipher Guedali's story. This reading relies on Frank's 2002 memoir, *At the Will of the Body: Reflections on Illness*, which he composed *post factum*--namely, after he recovered from testicular cancer. He narrates his experience of cancer in reverse, though reminding readers that this strategy is simply a stylistic excursus into his physical past and that his intention is to keep his mind and body in their present embodiment. Frank does not see any possibility for recovery without anchorage in the present, an idea that Guedali also intuits.

Frank, thus, prompts two puzzling ideas. From what we have noticed so far, both pain and recovery reside in the present, sometimes making it difficult to differentiate between the two. Frank is visited by uncertainty, too, when he cannot distinguish between what it is to be healthy and what it is to be ill. As he points out,

> [h]ealth and illness are not so different. In the best moments of my illnesses I have been most whole. In the worst moments of my health I am sick. Where should I live? [...] In 'health' there can only be fear of illness, and in 'illness' there is only discontent at not being healthy. In recovery I seek not health but a word that has no opposite, a word that just is, in itself. (135)

Frank gives here a carnal definition to a more and more forgotten, steeped-in-routine body; until a diagnosis, breakage, or rupture has occurred, we think of the body as just being "out there." However, even this vague idea could be symptomatic that something is wrong with the way in which we conceive of the body. Rather, a diagnosis, breakage, or rupture may allow us to acknowledge that the body traverses a stasis in which there is a delicate balance (if not an interchangeability) between concepts of health and illness, treatment and recovery. Etymologically, *convalescere* is a Latin word meaning "to grow strong." Invariably, growing strong and/or recuperating is done through resting. But resting in cases of illness often implies sleep and medication. As Frank writes, "I have learned that the changes that begin during illness do not end when treatment stops. [...] Those ill persons who recover must recover not only from the disease but also from being a patient" (97). His alternative, or compromise, is to think of (his) cancer as being in remission, where he oscillates between "periods of activity and treatment. [...] People in the remission society notice details more, because illness teaches the value as well as the danger of the everyday" (138-39). In so thinking, he admits that his body may break yet another time. Therefore, he emphasizes the crucial element that defines our lives, namely the inescapable randomness that sometimes ignores our plans, needs, and desires and challenges us to confront the breakage in our routine.

Undoubtedly, Guedali is not aware of Frank's concept, where recovery from an illness is perceived as remission. Nonetheless, like patients whose intuition is sharpened by the onset of an illness and treatment, Guedali senses that his remission could best be expressed through galloping. He knows that being confined to a hospital bed comes at the loss, even if temporarily, of bodily motility. After his surgery, then, he envisions himself as a centaur who runs free on the Brazialian pampas.

Au Lieu de la Conclusion

Kafka and Scliar demonstrate allegorically that matter is in continuous motion. There is one way to think of matter as transforming according to its inner structures, and thus not overstepping dangerously the threshold between the human and non-human, the normal and abnormal. Yet there is a totally different effect when matter is seen as transforming into something unpredictable and unrecognizable that defies common sense.

Within the range of exuberance and innocence to maturation and annihilation there are uncountable stages of fear derived from the entropic identity of our bodies. In these stories, we have noticed how two characters undergo severe physical and mental transformations. Could we suggest that, at some level, science itself has undergone a dramatic transformation? Has science evolved or involved? As Terry points out,

> [n]ow, in the late 20th century, after revelations about Nazi medicine, after invention of the nuclear bomb and germ warfare, and after the ecological horror wrought by technological developments, many of us have developed a very skeptical view of the doctrine of rationality and its practitioners. (137)

Furthermore, "We are living now in the age of the magical sign of the gene. [...] Scientists promise that if we can figure out the exact function and location of genes within the human body, the human population could be rid of diseases and defects" (Terry 148). We can infer that human beings are on the verge of crossing the boundary of being human, and are thus advancing toward hybridity. New shapes for our bodies have already been engineered and advertised in mass-media outlets--prosthetic devices, artificial hearts and kidneys, transplants, xenotransplants, cosmetic surgery, drugs. As suggested in *The Metamorphosis* and *The Centaur in the Garden*, and taking into account the technological delirium in which we live, to be *mise en corps* suggests a conglomeration of socially mediated altered and altering building blocks of physicality and technology, which adapt to each other either organically or by force.

While these works treat such transformations/mutations allegorically, on the other hand,

> [b]iotechnology applications involve the creation, combination, and construction of animals to contain the genetic traits of several species, including human being. Similar to a collage, transgenic animals are an aggregation of extant materials. Genes are reassembled and recombined to create more efficient and profitable 'super-animals' for food or pharmaceutical production, to develop research models for neurodegenerative diseases, and to expand further the sources of tissue and organs rendered available for human transplantation. (Anker 90)

Admittedly, today it is a challenge, if not an impossibility, to maintain an exclusively human identity throughout our lives. Our contours and shapes are more and more blurred. Based on the works discussed here, could we conclude by saying that a person-patient, who gradually loses his/her carnality, eventually becomes a body-without-objects? That is to say, person-patients may experience corporeal liminality not only from within as an undesired effect of their illness, be it mental or physical, but also from without as long as they cannot anchor their interest and perception in any object.

As Cataldi remarks, "[i]t is because of the percipient side of our flesh—this sensitive 'other side of our body' and its distancing from the depth of perceptible flesh—that we are 'incomplete,' as a riddle, gaping, open" (66). We did once venture to fail, and then answered correctly the Sphinx's riddle. Does she know how to answer our riddle? Perhaps we have been, and still are, dreaming of a body that has never been born; or, once born, does not know its limits and is, thus, eternal. Perhaps this happens precisely because of our recurrent dream of immortality where we have dared to defy our ontological status. As always, after briefly sojourning and disappearing into immortality, that is to say, after having committed mental hubris, we fall and fold back into ourselves, necessarily fabricated by imagination.

Notes

[1] The Cancer WEB Project, *Online Medical Dictionary,* Department of Medical Oncology, University of Newcastle-upon-Tyne, http://cancerweb.ncl.ac.uk/cgibin/omd?query=&action=Home, accessed December 16, 1997.

CONCLUSION

> We can endure almost any kind of suffering if we can grasp its meaning; if we can somehow connect the pain with some deeper significance.
> (Mansfield 18)

> The way that people can be spoken of
> Is not the constant way;
> The name that can be named
> Is not the constant name
> (Tzu 57)

> Knead clay in order to make a vessel. Adapt to nothing therein to the purpose in hand, and you will have the use of the vessel. Cut out doors and windows in order to make a room. Adapt to nothing therein to the purpose in hand, and you will have the use of the room.
> (Tzu 67)

In his *Synchronicity, Science, and Self-Making: Understanding Jungian Synchronicity through Physics, Buddhism, and Philosophy* (1995), Victor Mansfield attempts to explain some of our unconscious drives through an Jungian perspective, suggesting that we have an inborn capacity and desire to test the righteousness of our actions. This explains, at least in part, why we are determined to find the meaning of our actions, explain our behavior and learn a lesson of moral conduct. But does human suffering have ANY meaning? Could we ever understand pain and suffering as part of a collective datum? I am not arguing toward disregarding these two major, unavoidable conditions in everyone's life; however, it seems that they have been explained too much through collective theorems rather than allowing them (to have) the privacy and thus singularity each of them deserves. In trying to find their meaning, we inevitably become dominated by these two delicate moments of our lives and linger a tad too long on their significance. At the opposite end of this imaginary thread or axis, there is the ancient example given by Lao Tze in which he teaches us to let things and events flow and overflow our emotions, not with one but with uncountable meanings. In so doing, we do not attach to one definition and follow it narrowly, but we "adapt to nothing" and accept the unscripted scenario of our bodies and minds.

Furthermore, in trying to find a definition to pain and suffering, we are fixated on keeping the illusion of our wholeness, as if without this definition we would be impaired and fragmented, incapable of functioning. When the question is raised, "Where does it hurt?," it reveals more than a dialogue between a physician and a person/patient. There are others who (try to) define pain and suffering and realign them into some collective perspective on the human condition and what it means to be a living person. Today there is a definition for pain and suffering proposed by a physician, a nurse, a priest, a therapist, a friend and even a stranger. We have actually created specialized therapists—the algologists—whom we ask to mediate our (understanding of) pain. It appears that when in severe physical discomfort/breakage, we need to have as many perspectives on it and definitions as possible, to have others negotiate for us this delicate encounter.

What is even more troublesome is that we try to control our emotions. We are hyperactive during the day (helped by caffeinated beverages and/or other stimulants meant-to-boost energy levels), but we would like to be tranquil during the night. We want to live the lives of more than one person during the day, but somehow return to quietude and effortless sleep during the night. We have already intervened in our health structure by exposing ourselves physically and emotionally to an abnormal rhythm full of all sorts of requirements. We try to control our minds and bodies chemically because we do not accept that our lives may be interrupted by pain and suffering. What concerns me, then, is how much we ask of our bodies and minds and yet are still surprised at how often they break into afflictions that typically do not necessitate hospitalization and are (in many cases) improperly treated with over-the-counter drugs (e.g., depression, eating and sleeping disorders, chronic fatigue, etc.). According to Emily Martin,

> The original meaning of the word 'immune' was to be exempt from the requirement of service to the state. It was not until the late 19th-century that the entity which the 'immune' individual could ward off became not the state, but disease. The two meanings of the word came together when immunization was developed and brought to us, often without choice, through resources and personnel mobilized by the state. (132)

Furthermore, because we cannot accept pain, which we think rightly to be degrading, we sometimes dream of finding a method that could eradicate pain altogether from our lives.

Because we live in a culture in which pain is often stigmatized as shameful, we have a tendency to make sure we do not become its "victims." This idea is not novel. Ever since the eugenic movement, an emphasis has been maintained on the dichotomy between ill and healthy that operates on the exclusion principle: whoever is not healthy is not able to function in, and, hence, is excluded from, society. As Gilman argues, "It is not only that the healthy becomes the beautiful, but that the beautiful becomes the healthy; the diseased is not only the ugly, but the ugly, the diseased. And the ugly must be made to give way to the beautiful through the agency of scientific medicine" (51). Granted, these ideas appeared at the beginning of the 20th-century, when concepts such as social disease were coined. During those times, there emerged social medicine, the branch created to take control over the "ugly" (meaning "not functional"). Lupton reminds us that "In the closing years of the 19th-century, a new regime of public health developed [due to] the discovery of microbes. [...] The 'enemy' [the microbe] had now been rendered visible, and a logical sequence of events could be outlined to link the causes of illness" (*The Imperative* 36). Arguably, the most evident reason for this theory was, in fact, a critique addressed to the citizen. According to these theories, someone was regarded/accepted as a citizen if s/he was fully responsible for his/her own health and maintained an appropriate lifestyle by controlling his/her diet, exercising, and becoming fit.

Soon a whole industry emerged in which people were bombarded by this powerful, often abusive, campaign over how to became a healthy and thus responsible citizen. Patent medicines become available on the market. In *The Consuming Body* (1994), Falk remarks that

> By the end of the 19th-century all patent medicines had redefined themselves not only as a cure for the sick and frail, but also as a preventive elixir for the healthy. A patent medicine called Pure Malt Nectar, for instance, was advertised under this slogan in the United States: 'invigorating tonic, alike for invalids and those in health.' [...] Preventive medicines promised to keep the evil (illness, ageing and ultimately death) away and turn the 'normal' life into a better one by providing extra energy, happiness, beauty, etc. The same arguments were used also in the marketing of cornflakes, rice krispies, etc., in the name of 'good health' (167).

On one hand, there emerged the theory about microbes; on the other hand, there are vitamins and other supplements to keep us "healthy" and in shape. Even today, the emphasis is still placed on these two, disregarding pollution, stress, inadequate treatments and even the transgenic biotechnological experiments. Health seems to be simplistically reduced to an equation of eating right, keeping oneself clean (and away from those who are not), and exercising. All these practices are effective at maintaining a certain image, but they do not necessarily address health-related issues.

This desire to keep us whole and healthy is, in part, a rereading of the Human Genome Project. Undeniably as with many achievements, there is a false hope within this project, too. If patent medicines promise to keep both "invalids" and those healthy in shape, if these cure-alls are targeted to both parties, the Human Genome Project moves its fantasies a step further, where discovering the "wrong" gene within one body may make him/her better control his/her pain, and eventually find ways to eradicate it. I cannot honestly imagine such a type of population; would it be perfect? What would it be its purpose? If such a population would have neither diseases nor defects, would it live permanently and be happy? One could only speculate on finding answers to these polemical, if not unpractical, questions. Then, because those who had already been born could not be helped to rid themselves of pain, suffering and the humiliation involved in both, we may yet have to think of an alternative, compromising solution to explain why we cannot permanently keep ourselves healthy. In his essay "The Ethics and Politics of Caring: Postmodern Reflections," Nick Fox notes that

> In their political philosophy of resistance, Deleuze and Guattari provide us with the anti-icon for a postmodern ethos: the *nomad*. [...] The nomad does not put down roots, or manipulate her environment to suit her needs and wishes. She does not seek control, she takes what is on offer, assimilates it, and moves on. She refuses to acknowledge the forces which would territorialise her, the rationalism which values the stable [...]. It is hard to be a nomad; in fact, there *are* no nomads, there is only nomadism, it is a process, not an identity. Nomadism is about becoming other, and one never finally becomes other, rather, we lurch from one identity to another. (339-40)

It is plausible to argue that nomadism is a counter-rhetoric to the previously mentioned health-related campaigns, which addressed the urgency of keeping oneself healthy through a certain type of self-control. On the contrary, nomadism admits that it is only theoretically effective, that, as long as there is no nomad, there is no stable and uninterrupted identity for anyone living, that identity is a process which involves motion and erosions, breaks and attempted repairs. Put differently, nomadism may be effective in health-related campaigns because it suggests that one may have an identity as long as one is willing to admit that that identity is not definitive and is, in fact, breakable. By the same token, one may possess a good health status as long as one understands that that status is exposed to many unpredictable changes.

However, in a society that promotes patent medicines and health campaigns about responsible citizens, could nomadism ever work? In *The Shattered Self: The End of Natural Evolution* (2001), Baldi proposes his view about the old theory of our gradually decentering from—and I would add re-centering to—the center of the universe:

> In ancient times we thought we were made of a stuff completely different from, say, rocks, or that we were very different from other animals or plants, and that we occupied a special position in nature. [Today we know that] *homo sapiens* is made of atoms, like all other macroscopic objects we know. […] [Yet,] At the physchological level, in particular, we view ourselves […] as a special and unique entity endowed with special feelings, intelligence and thoughts. (10-12)

To better understand both nomadism and decentering, it helps to note that in previous times, the suffix "–ism" was used to define as clearly as possible a moment in history and mark an important cultural episode in human evolution (e.g., Classicism, Romanticism, and even the highly ambiguous term Postmodernism). Now words try to group us together: nomad-ism—where one does not have a stable identity—and health-ism where "the maintenance of good health is the responsibility of the individual" (Lupton, *The Imperative* 70). Undeniably, health is a vexing notion, and it is mediated and altered by culture, socio-political discourses, and the attendees of those in pain; but, in the end, responses to illnesses reflect a stoic test of personal endurance. An illness may have the potential to "decenter" one person/patient from his/her former identity, yet eventually this experience turns out to be a recentering and discovery of one's ignored self. An illness, no matter how traumatic, is a non-removable part of one person's/patient's status.

Therefore, like happiness, the word health, has been overused until it has begun to have a highly debatable and controversial meaning. Carol Donley and Sheryl Buckley note that

> many terms [that] are used to describe people outside physiological norms are negative and stigmatizing—freaks, mutants, monsters, mistakes of nature [again subtle allusions to eugenic readings]. Clinical terminology, such as *terata*, seems euphemistic. And terms such as *abnormal*, *disabled*, *malformed*, and *deviant* begin with a negative prefix that puts the person being described in opposition with the normal, the able, the formed (171).

In the passage that preceded this remark, the authors had introduced their view about the Procustean legend according to which the monster's victims needed to fit into his iron bed, otherwise their bodies were subjected to violence (either stretched when they were too small, or chopped off when they were too long). In this light, the abnormal, disabled, malformed and deviant are placed more emphatically next to the normal, the able, and the formed as the two distinct theoretical discourses through which we tend to define people and events. However, no human being should ever be expected to fit into any ideological Procustean bed, regardless of how much we want to eradicate pain and suffering.

Unfortunately, as noticed through the primary works used in this study, life is completely changed/altered when one person/patient cannot move, lies down in bed for day after day, takes several medications and barely engages in social activities. When pain becomes a constant, life on the horizontal may look as if one person/patient has lost his/her spine. However, the first part of my conclusion does not focus on those who are severely ill but proposes to bring to attention the emergence of a new type of liminal embodiment today or *le corps déjà-vu*. This type of body seems to be under constant external pressure as well as a self-imposed desire to improve itself, thus losing its personality/uniqueness and fitting more or less into an advertised standard of beauty (and social functionality). Having been exposed to several stories, visual ads and

socio-medical discourses about the suffering body, *le corps déjà-vu* tries desperately to keep itself safe and healthy by an intake of pills and supplements, products that promise to keep its structure intact by frugal diet and excessive exercise. The new physical liminality of an alarmingly *still* healthy body requires elaborate research, which at the moment does not constitute a part of this work.

These generalities aside, let me return to the main interest of this study. I started by stating that it was not about my suffering and pain, but only a "travelogue" into others' experience of physical and emotional decline. In reaching its conclusion, I do not undergo a transference syndrome claiming now that this has somehow become a study about my pain and suffering, but I do admit that it has affected me tremendously. *If* I have not been seriously ill and hospitalized, *if* I have been able to more or less maintain my healthy status, does it mean that I know what it is to be healthy (and this is how I introduce the conditional ambiguity of my own frustration vis-à-vis pain)? Throughout the process of writing this book (when I was also raising my baby boy), I have experienced episodes of chronic fatigue and, in order to continue, I had to indulge myself in moments of rest and relaxation. About this type of pain and suffering, Morris believes that it makes the "[p]atients […] move in an in-between realm: they clearly are not well, but their malady will not let us see them as absolutely sick, An affliction that operates in-between and in secret, of course, generates endless paradoxes" (67).

Are we a culture of pain or of health because it appears that we are in dire need of constantly repairing and/or improving ourselves even when no such actions are required? What does it mean to be well? Do minor discomforts disrupt our wellness? Or, on the contrary, do only major breakages make us think about pain? Equally important to ask is, how many times is the word health used? Undertake an experiment, and for a week keep a rudimentary journal where you note how often you encounter this word: on TV, in newspapers, in ads, in products that you buy and in daily conversations. In one recent advertisement, there is apparently a chair so craftily designed that it claims to maintain the sitter's "good" health. Apart from the outrageous claim, is there such a thing as "good" health? Also, there exist on the market today wipes which, as one reads on their label, can prevent people from getting the flu virus. These two of many uncountable products seem to mock the buyer's idea of what is to be healthy rather than to propose an effective education and/or help. They somehow send me back to the slogan of Vogel's play, "Don't sit, do squat!" Here "don't sit" could be decoded as: "Do not be impassive, keep buying products and invest in the "good" health (or should I say image?) of your body.

In the concluding section of this study, I want to reflect upon the nature of liminality via Gunther von Hagens' sculptures as part of his *Body Worlds* (2005) (Initially, these works were supposed to be the subject of an ample ending discussion). Right now, I frankly admit, after I have introduced such vibrant confessional stories, I doubt these sculptures could serve as a proper, effective example for my conclusion. I am not sure, in fact, if they belong in a museum or if they should be put back in formol and stored in a lab; if they are indeed pedagogically-oriented, as their artist envisions them to be, or if they somehow test/defy our visual empathy? (As a parenthetical observation, these sculptures are exhibited to the public in various postures: one sits at the table playing chess, another appears to be in motion and kicks a football, yet another rides a horse—all creating the illusion they are still endowed with human attributes). These sculptures are real human beings whose dead bodies have been preserved through the plastination technique. Around the 70's, while still a resident at the University of Heildeberg, Dr. von Hagens focused on refining the plastination technique; by replacing a dead body's fluids and fats with plastic, the resulting works are then exposed to light, heat or gas to maintain their lifelike appearance. On seeing them, viewers may want to approach them from a closer angle, but at the same time von Hagens' works seem to warn: *Noli me tangere* (i.e., "Do not touch me!"). That may be, in fact, the moment when one realizes that they have been violently and unethically flayed (that is the general impression, at least).

Imagine that if the images generated by the VHP were considered controversial and, "Once a cadaver, [...] recognized as a living subject, is thus encoded and stored in a database, then the dead are no longer simply the not living" (Thacher 176-77), then how much more discomfort could von Hagens' work bring to the viewer? Their all-enveloping red cover/surface burns my retina, and, without realizing it, I immediately placed them in my mental visual archive next to Chaim Soutine's 1925 painting *Carcass of Beef*, of which art historian Richard Leppert writes: "The carcass confronts us in a close-up, in-your-face manner, hung from spirals of rope and stretched open, freshly bleeding. [...] The carcass appears to have been quickly gutted and skinned [and it] reveals something near to desperation" (94). Then, for the artist, "[t]he ox's death is pointless except as a demonstration of violence and the potential of living things to suffer" (Leppert 95). For Soutine the project of painting several carcasses revealed the unnecessary violence to which the animals had been subjected, and thus exposed the greediness through which we want to mark our ostentatious supremacy among living creatures.

On the other hand, when von Hagens' sculptures (/specimens) are put next to the primary works analyzed in my study, they do not seem to enter into a vibrant dialog with them; instead, their greatest achievement is to shock us without offering any comfort and/or explanation.

Therefore, I have decided that von Hagens' sculptures are not really an appropriate focus for the concluding section because my main interest throughout this project has been directed toward *homo patens/patendus* (i.e., "the suffering/*about to* suffer being"). How much do we know about the pain and suffering of the *living* other? With a slight difference, this question was raised by Robert Lincoln Watkins, a New York physician, in a publication back in1902: "'What can you tell about a man without seeing him, say, from an examination of the photo'?" (Cartwright 81). My book is a reading of the persons'/patients' stories that I discovered more or less by accident. I structured this study in six chapters, but rereading them, I realize that I could have dedicated to each one of these "stories" a special chapter. In addition, their experience of pain is so profound that they could easily become part of different chapters. For example, I could have used Butler's participation in Rosenblum's pain as one of the examples within the chapter entitled "At the Edge: The Other's Liminality" (and/or say something about her cancer, diagnosed after the death of her life partner, even though this occupies a marginal space in the memoir). I could have used Gonzales-Torres's works as examples of a "body-broken" or a "reclaimed body" since he had, and eventually died because of, AIDS. I felt the need, finally, to exclude von Hagens' works, but I added something that was not initially part of this study, namely Silver's short story about a mother and a daughter sharing a train compartment heading to a clinic of ambiguous destination, of either hope or of agony.

More books could be added, such as: Jean Pond's *Surviving* (a memoir about surviving a brain tumor), Robert F. Murphy's *The Body Silent* (a memoir about a paralyzed man), Paul Monette's *Borrowed Time: An AIDS Memoir* (one of the first memoirs written on the subject of AIDS), Jean-Dominique Bauby's *The Diving Bell and the Butterfly* (a memoir about the "lock-in syndrome" of a man who suffered a severe stroke), Joan Didion's *The Year of Magical Thinking* (a book that urges the importance of medical education as an imperative civic act), Randy Pausch's *The Last Lecture* (about an academic who does not to give up when facing the "deadline" of his illness), as well as other works, both textual and visual. The concept of liminality via a body severely collapsed could be expanded to other persons/patients who have been suffering from other equally depriving illnesses, such as Alzheimer's (for example, Sarah Polley's film *Away from Her*). By the same token, a recent book about the invention of ailments could be helpful (Joerg Blech's *Inventing Disease and Pushing Pills: Pharmaceutical Companies and the Medicalisaion of Normal Life*). But if I feel inclined to strongly add anything, that would be a study involving either persons/patients recently returned from hospitalization or one conducted in the hospital, the epicenter of fear and despair for both persons/patients, their attendees, and medical personnel. That may help better parallel the differences in the perception of pain. In this study I used only those stories that have circulated almost exclusively in academia, so they reflect a certain type of ideology and writing. Still, I wrote it not only because I wanted to share different perspectives on the subject of liminality as foregrounded by acute pain and suffering, but also because I think the academic curriculum embraces them almost exclusively from a close reading perspective at best, rather than revealing the social undercurrent of these works.

In this light, it is useful to emphasize that pathology is the study of suffering (as generated by illnesses). Could we introduce it as part of the curriculum in the Humanities? Could we teach literature and art from that perspective? Skeptics who have been trained in canonical thinking say we should not. I beg to differ. If students in medicine are exposed to literary texts in order to familiarize themselves with literary and thus socio-cultural concepts about pain and suffering, why aren't we exposed to medical treatises, no matter how rudimentary? Don't we all need to know our bodies, prior to their seriously being affected *and* transformed by pain? I have already suggested some answers to these questions in my study, but I am still confused by what it means to live a life in agonizing pain. In one of her novels, Nadine Gordimer writes sarcastically, "When in doubt go to the dictionary" (47). When in doubt about our physical discomfort, could we rely on the other's confession of pain? Could we return to the body-in-pain, either by reading its story or examining it visually? In 2002, Pat Croskerry--a physician-- wrote an article, "Achieving Quality in Clinical Decision Making: Cognitive Strategies and Detection of Bias," where he stated very frankly that doctors have a tendency to diagnose their patients after a couple of routine and rather informal dialogue exchanges and minimal check-ups. Croskerry explained or justified this bias in thinking by saying that doctors have a template of diseases, just as, comparatively, literati know the difference between metaphor and synecdoche, mathematicians the difference between positive and negative numbers, and art historians the difference between Cubism and Fauvism.[1] Could literati err and confound a metaphor as a synecdoche? Undoubtedly that has happened and will continue to happen because *errare humanum est* (i.e., "It is human/acceptable to err"), as the old saying goes.

But what can one person/patient do when a doctor misreads his/her diagnosis? By trial and error, more and more doctors (must) admit today that the best, or most decent treatment, is to tell/inform their patients that what they are about to prescribe may not always be efficient. As Michael Bury notes, "[t]oday diagnostic is often probabilistic rather than definitive. Treatment, in turn, may often be 'palliative,' that is trying to reduce the impact of symptoms, or contain the disease, rather than hoping to cure it completely" (5). Unfortunately, based on the testimonies used in this study, persons/patients do not always expect that to be explicitly disclosed during an appointment with a doctor. This is only what some of us know, thus being able to better address our concerns with our doctors. Others have not been exposed to this kind of knowledge and still invest doctors with more power over their bodies than they should have. This is one more urgent reason why it is important to introduce pathology into the curriculum, because the human body is a highly heuristic site where different discourses collapse, but where we have the moral obligation to not let doctors give us their perhaps too quick, formulaic diagnosis--understanding by this that, in most of the cases, signs of an ulterior bodily breakage are evident even before our appointments with doctors.

We are a mediated and medicated society. Lupton believes that "Every individual is now involved in observing, imposing and enforcing the regulations of public health, particularly through the techniques of self-surveillance and bodily encouragement by the imperatives of health promotion" (*The Imperative* 76). However, this imperative is not as blatantly expressed in the works analyzed throughout this study. On the contrary, each person/patient exposed his/her fear *vis-à-vis* insecure or improper diagnosis, or poor/deficient medical insurance, or treatments with side-effects and unpredictable emotional chain reactions. They had difficulties over accepting this altered embodiment developing at rapid speed and exhibiting signs of severe decline. Hence, apart from their attempts at recovery, they faced a problem which openly confronted their intimacy. We do not always manage to deal with a "body-broken," a swept away former perception of the self, and a delicate negotiation with a new identity. We live in a closed or daily ignored body. As Kurosawa's Watanabe showed us, an illness opens it up, but by that time, this opening up may have the impact of an avalanche. Donna Haraway says that "Life is a window of vulnerability. It seems a mistake to close it" (224). A window accommodates shades and patches of undiluted light; it is left ajar or tightly closed; it lets the air circulate or it makes it stale; finally, it permits a dialogue with the outside or it lets the walls do the talking.

In addition, life as a window of vulnerability is thus always liminal, yet ambiguous. I started this book wanting to find a definition for what it means to be ill. In reaching its conclusion, I find it more troublesome to define what it means to be (and feel) healthy. Maybe the well body is actually never born; maybe it is a creation of the collective unconscious or a strong, unfulfilled desire that teases us now and then. Happiness and wellness are tantalizing at the beginning of the 21st-century as long as we cannot deny our inevitable encounter with pain. In this context, this study cannot exclusively remain a "travelogue" anymore. If life is a window to vulnerability, life is a window to *vulnus* or "wound" (and we all get wounded at some point in our lives). The body has a tremendous capacity to restart itself after a break. Within us there is a network of "backup" systems, and if one fails, one is replaced: "We have an extra kidney, an extra lung, an extra gonad, extra teeth. The DNA in our cells is frequently damaged under routine conditions, but our cells have a number of DNA repair systems. If a key gene is permanently damaged, there are usually extra copies of the gene nearby" (Gawande 52). In other words, the body speaks a langue of adaptability, in which each of us has his/her own parole. Within us, as a character in Lavery's play remarked, there is the magnificent "Human Will to Live." There is also, as Gonzales-Torres would argue, a passion to (be) continue(d).

What this study has done is to look at persons/patients who realized they might have had a better way of coping with their abrupt decline if they possessed a minimal knowledge of anatomy and if they were more intimate with their bodies. In this context, a good example comes from Spence's works, which showed that photographs of ill persons disclose another level of intimacy that does not always have to happen behind closed doors, but could also (and, admittedly, more potently) occur in public—taking into consideration her artistic vision of art as phototherapy that focuses on naked and wounded bodies. By so doing, she reinforced her belief that the body in pain should not be alienated from its traumatic experience, but instead should be kept visible in the audience's focus. Intimacy, then, may be itself a *wound* that asserts all (the wanted and especially unwanted) transformations that our bodies undergo through time. The type of intimacy envisioned by Spence may thus serve as a means to alleviate our perceptions of a body that is constantly changing, whether or not it has inscribed on it (permanent) signs of illness.

I end this study by referring to a painting dating from around 1551 by Jacopo Bassano, *Saints Sebastian and Roch*. It is an oil on canvas depicting the two saints. Pairing them is quite surprising to the contemporary viewer: the violently- pierced-with-darts martyr is placed next to a severely ill one. About this work, Catherine Wilcox-Titus remarks that "St. Roch looks directly at the viewer as he points to his thigh, a decorous way to indicate the swelling and soreness of the lymph tissue in his groin typical of the symptoms of Plague. This work is actually a banner used as an ex voto, meaning the work was commissioned by a grateful patron who had survived the disease." It also helps to note that this painting was recently part of the exhibit *Hope and Healing: Painting in Italy in a Time of Plague, 1500-1800*. The pathos of all the paintings is impressive considering the artists' effort and will to leave as testimony their struggle with pain, and in the end--cathartically--consider it to stand as a moral example. About this exhibit, Holland Carter notes that during those times, "Devotion alone […] was not always enough. You said your prayers, and the plague raged on. So some people pursued the more proactive, practical option of pious deeds. And no deeds were more usefully humane than the so-called corporal acts of mercy (feeding the hungry, caring for the sick and burying the dead)" (27). I use this image as the visual epigraph for my conclusion because it represents a lesson of the suffering body as well as of its unspoken, yet accepted humility.

Today we have not exactly become separated from parabolic readings for our bodies; nonetheless, we have moved toward improving our bodies and hence sometimes drastically intervene in their internal constitution. The twisted, coiled-on-the-rod serpent has stood as the symbol for medical art and practice for centuries. With its wrapped-like structure or pose ready at any point to be branching out in different and unpredictable directions, the serpent makes me think of the helical structure of our DNA (which encodes our essence/singularity), as well as the combinatory yet surprising dimensions of health and illness that are coiled inside our minds and bodies. Then, maybe, we realize that both health and illness are irreversibly liminal and dual states, and that it may not be too far-fetched to envision the serpent as a representative imagine for our becoming persons/patients.

Notes

[1] Croskerry identified several biases: 1. "anchoring bias," where "Combinations of salient features of a presentation often result in pattern recognition of a specific disease" (1186); 2. "confirmation bias" when accumulated "data can be selectively marshaled to support a favored hypothesis" (1186); and 3. "value-induced bias" when "any biasing tendency toward worst-case scenarios increases the likelihood of detection of diagnosis that must not be missed" (1186).

REFERENCES

Abram, David. *The Spell of the Sensuous: Perception and Language in a More-than-Human World*. New York: Vintage Books, 1997.

Almodóvar, Pedro. *Talk to Her*. 2002.

—. *All about My Mother*. 1999.

Amenábar, Alejandro. *The Sea Inside*. 2004.

Anker, Suzanne and Dorothy Nelkin. *The Molecular Gaze: Art in the Genetic Age*. Cold Spring Harbor: Cold Spring Harbor Laboratory Press, 2004.

Babenco, Hector. *Carandiru*. 2003.

Bachelard, Gaston. *The Poetics of Space*. Trans. Maria Jolas. Boston: Beacon Press, 1969.

Barthes, Roland. *Image, Music, Text*. New York: Hill and Wang, 1977.

Baldi, Pierre. *The Shattered Self: The End of Natural Evolution*. Cambridge: MIT, 2001.

Bautz, Lewis. "(Sans Titre), Félix Gonzales-Torres." *Félix Gonzales-Torres*. Ed. Julie Ault. Göttingen: SteidlDangin, 2006: 211-216.

Beckett, Samuel. *Krapp's Last Tape and Other Dramatic Pieces*. New York: Grove Press. Inc., 1957.

Berger, John. "Understanding a Photograph." *Classic Essays on Photography*. Ed. Alan Trachtenberg. New Haven: Leete's Island Books, 1980: 291-294.

Berman, Morris. *Reenchantment of the World*. Ithica: Cornell UP, 1981.

Bennett, Jane. *The Enchantment of Modern Life: Attachments, Crossings, and Ethics*. Princeton: Princeton UP, 2001.

Bordo, Susan R. *The Flight from Objectivity: Essays on Cartesianism and Culture*. Albany: State University of New York Press, 1997.

Bury, Michael. *Health and Illness in a Changing Society*. New York: Routledge, 1997.

Cameron, Dan. *Fever: The Art of David Wojnarowicz*. New York: Rizzoli, 1998.

Canning, Peter. "The Brain Is the Screen: An Interview with Gilles Deleuze." *The Brain Is the Screen: Deleuze and the Philosophy of Cinema*. Ed.Gregory Flaxman. Minneapolis: University of Minnesota Press, 2000: 365-373.

Carson, Roland A. "The Hyphened Space: Liminality in the Doctor-Patient Relationship." *Stories Matter: The Role of Narrative in Medical Ethics*. Ed. Rita Charon. New York: Routledge, 2002: 171-181.

Cassuto, Leonard. "Oliver Sacks and the Medical Case Narrative." *Disability Studies: Enabling the Humanities*. Ed. Sharon L. Snyder. New York: MLA, 2002:118-131.

Carter, Holland. "Desperately Painting the Plague." *The New York Times*. 29 July (2005): 25-27.

Carter, Thatcher. "Body Count: Autobiographies by Women Living with Breast Cancer." *JPC*. 36.4 (2003): 653- 668.

Cartwright, Lisa. *Screening the Body: Tracing Medicine's Visual Culture*. Minneapolis: University of Minnesota Press, 1995.

Cataldi, Sue E. *Emotion, Depth and Flesh: A Study of Sensitive Space: Reflections upon Merleau-Ponty's Philosophy of Embodiment*. Albany: State University of New York Press, 1993.

Caruth, Cathy. *Unclaimed Experience : Trauma, Narrative, and History*. Baltimore: Johns Hopkins University Press, 1996.

Chambers, Tom. "What to Expect from an Ethics Case." *Stories and their Limits: Narrative Approaches to Bioethics*. Ed. Hilde Lindemann Nelson. New York: Routledge, 1997: 171-184.

Clark, Brian. *Whose Life Is It Anyway?* New York: Mead, 1979.

Connor, Steven. "Mortification." *Thinking through the Skin*. Eds. Sara Ahmed and Jackie Stacey. New York: Routledge, 2001: 35-51.

Croskerry, Pat. "Achieving Quality in Clinical Decision Making: Cognitive Strategies and Detection of Bias." *Academic Emergency Medicine: Official Journal of the Society for Academic Emergency Medicine*. 9 (2002): 1184-1204.

Cubitt, Sean. *The Cinema Effect*. Cambridge: MIT Press, 2004.

Darwish, Mahmoud. "Remainder of a Life." Trans. Fady Joudah. *The New Yorker*. 14 (2007): 119.

Deleuze, Gilles and Félix Guattari. *A Thousand Plateaus: Capitalism and Schizophrenia*. Minneapolis: University of Minnesota Press, 1987.

Donley, Carol and Sheryl Buckley. "The Tyranny of the Normal." *Teaching Literature and Medicine*. Eds. Anne Hunsaker Hawkins and Marilyn Chandler McEntyre. New York: MLA, 2000: 163-174.

Dossey, Larry. *Space, Time, and Medicine*. Boulder: Shambhala Publications, Inc., 1982.

D'Lugo, Marvin. *Pedro Almodóvar*. Urbana: University of Illinois Press, 2006.

Eberwein, Robert T. *Film & the Dream Screen: A Sleep and a Forgetting*. Princeton: Princeton UP, 1984.
Edson, Margaret. *Wit*. New York: Faber and Faber, 1999.
Elias, Norbert. *The Loneliness of Dying*. New York: Blackwell, 1985.
Falk, Pasi. *The Consuming Body*. Thousand Oaks: Sage Publications, 1994.
Ferguson, Russell. "Authority Figure." *Félix Gonzales-Torres*. Ed. Julie Ault. Göttingen: SteidlDangin, 2006: 81-104.
Flaxman, Gregory. *The Brain is the Screen: Deleuze and the Philosophy of Cinema*. Minneapolis: University of Minnesota Press, 2000.
Frank, Arthur W. *At the Will of the Body: Reflections on Illness*. Boston: Houghton Mifflin Company, 2002.
Foster, Hal. *Prosthetic Gods*. Cambridge: MIT Press, 2004.
Foucault, Michel. *Discipline and Punish: The Birth of the Prison*. Trans. Alan Sheridan. New York: Pantheon Books, 1977.
—. *History of Sexuality. The Use of Pleasure*. Trans. Robert Hurley. New York: Pantheon Books, 1985.
—. *The Birth of the Clinic: An Archeology of Medical Perception*. Trans. A. M. Sheridan Smith. New York: Pantheon Books, 1973.
Fox, Nick. "The Ethics and Politics of Caring: Postmodern Reflections." *Health, Medicine and Society: Key Theories, Future Agendas*. Eds. Simon J Williams, Jonathan Gabe and Michael Calnan. New York: Routledge, 333-349.
Fox, Renée. "Through the Lenses of Biology and Sociology: Organ Replacement." *Debating Biology: Sociological Reflections on Health, Medicine, and Society*. Eds. Simon J. Williams, Lynda Birke, and Gillian A. Bendelow. New York: Routledge, 2003: 235-244.
Gadamer, Hans-Georg. *The Enigma of Health: The Art of Healing in a Scientific Age*. Trans. Jason Geiger and Nick Walker. Stanford: Stanford UP, 1996.
Gilman, Sander. *Picturing Health and Illness: Images of Identity and Difference*. Baltimore: Johns Hopkins UP, 1995.
Goldin, Nan. *Witnesses: Against our Vanishing*. New York: Artists Space, 1989.
Greenfield, Susan. "Inner Space." *Space in Science, Art, and Society*. François Penz,
Gregory Radick and Robert Howell, Eds. Cambridge: Cambridge UP, 2005: 6-21.
Grosz, Elizabeth. "Histories of the Present and Future: Feminism, Power, Bodies." *Thinking the Limits of the Body. Jeffrey Jerome Cohen and*

Gail Weiss, Eds. Albany: State University of New York Press, 2003: 13-23.
—. *Space, Time, and Perversion : Essays on the Politics of Bodies*. New York: Routledge, 1995.
Halberstam, Judith and Ira Livingston. "Introduction: Poshuman Bodies." *Posthuman Bodies*. Eds. Judith Halberstam and Ira Livingston. Bloomington: Indiana UP, 1995: 1-22.
Hallan, Elizabeth, Jenny Hockey and Glennys Howarth. "The Body in Death." *Reframing the Body*. Eds. Nick Watson and *Sarah Cunningham-Burley* New York: Palgrave, 2001: 63-77.
Haraway, Donna J. *Simians, Cyborgs, and Women: The Reinvention of Nature*. New York: Routledge, 1991.
Hawkins, Anne Hunsaker. *Reconstructing Illness: Studies in Pathography*. West Lafayette: Purdue UP, 1993.
Hoffman, Amy. *Hospital Time*. Durham: Duke UP, 1997.
Hoffman, William. *As Is*. New York: Random House, 1985.
Heidegger, Martin. *The Question Concerning Technology, and Other Essays*. Trans. William Lovitt. New York: Garland Pub., 1977.
Kafka, Franz. *The Metamorphosis*. Trans. Translation by Willa and Edwin Muir. New York: Schocken Books, 1968.
Kimmich, Allison. "Writing the Body: From Abject to Subject." *Auto/biography Studies*. 13. 2 (1998): 223-234.
Kovács, András Bálint. "The Film History of Thought." *The Brain is the Screen: Deleuze and the Philosophy of Cinema*. Ed. Gregory Flaxman. Minneapolis: University of Minnesota Press, 2000: 153-170.
Kramer, Larry. *The Normal Heart*. New York: New American Library, 1985.
—. *The Tragedy of Today's Gay*. New York: Penguin Group, 2005.
Krieg, Joann. *Epidemics in the Modern World*. New York: Maxwell Macmillan International, 1992.
Kushner, Tony. *Angels in America. Perestroika*. New York: Theatre Communications Group, 1994.
Kurosawa, Akira. *Ikiru*. 1952.
Kwon, Miwon. "The Becoming of a Work of Art: Félix Gonzales-Torres and a Possibility of Renewal, a Chance to Have a Fragile Truce." *Félix Gonzales-Torres*. Ed. Julie Ault. Göttingen: SteidlDangin, 2006: 281-314.
Laureyssens, Dirk. 2005. "The Big Tube Theory."
<http://www.mu6.com/pelastration.html>

Lavery, Bryony. Last Easter. NOT YET PUBLISHED. Available in photocopied manuscript only, courtesy of Dramatists Play Service, Inc. 2004.

Lawton, Julia. *Dying Process: Patients' Experiences of Palliative Care*. New York: Routledge, 2000.

Leder, Drew. *The Absent Body*. Chicago: University of Chicago Press, 1990.

Lefebvre, Henri. *The Production of Space*. Trans. Donald Nicholson-Smith. Cambridge: Blackwell, 1991.

Lippit, Akira Mizuta. "Phenomenologies of the Surface; Radiation-Body-Image." *Qui Parle*. 9.2 (1996): 31-50.

Lorde, Audre. *The Cancer Journals*. Argyle, New York: Spinsters Ink, 1980.

Lupton, Deborah. *The Imperative of Health: Public Health and the Regulated Body*. Thousand Oaks, California: Sage Publications, 1995.

—. *Medicine as Culture: Illness, Disease and the Body in Western Societies*. Thousand Oaks, California: Sage Publications, 2004.

Mansfield, Victor. *Synchronicity, Science, and Self-Making: Understanding Jungian Synchronicity through Physics, Buddhism, and Philosophy*. Chicago: Open Court, 1995.

Martin, Emily. "Flexible Bodies: Science and a New Culture of Health in the United States." *Health, Medicine and Society: Key Theories, Future Agendas*. Eds. Simon J. Williams, Jonathan Gabe and Michael Calnan. New York: Routledge, 2000: 125-145.

Martin, Jean-Clet. "'Of Images and Worlds: Toward a Geology of the Cinema.'" *The Brain is the Screen: Deleuze and the Philosophy of Cinema*. Ed. Gregory Flaxman. Minneapolis: University of Minnesota Press, 2000: 61-87.

McDonald, David. "Derrida and Pirandello: A Poststructuralist Analysis of *Six Characters in Search of an Author*." *Modern Drama*. 4.2 (1977): 412-435.

Merleau-Ponty, Maurice. *Phenomenology of Perception*. Trans. Collin Smith. New York: Routledge, 2002.

Merrell, Floyd. *Sensing Corporeally: Toward a Posthuman Understanding*. Toronto: University of Toronto Press, 2003.

Morris, David B. *The Culture of Pain*. Berkeley: University of California Press, 1991.

Nietzsche, Friedrich. *The Will to Power*. Trans. Walter Kaufmann and R. J. Hollingdale. New York: Vintage Books, 1968.

O'Neill, John. "Horror Autotoxicus: The Dual Economy of AIDS." *Contested Bodies*. Eds. Ruth Holliday and John Hassard. New York: Routledge, 2001: 179-186.

Onwurah, Ngozi. *The Body Beautiful*. 2001.

Perl, Sondra. Felt Sense: *Writing with the Body*. Portsmouth: Cook Heinemann, 2004.

Phelan, Peggy. *Mourning Sex: Performing Public Memories*. New York: Routledge, 1997.

Pinter, Harold. "Art, Truth, and Politics." *PMLA*. 21(2006): 805-818.

—. *Other Places: Three Plays*. London: Methuen, 1982.

Plantinga, Carl. "The Scene of Empathy and the Human Face on Film." *Passionate views: Film, Cognition, and Emotion*. Eds. Carl Plantinga and Greg M. Smith. Baltimore: Johns Hopkins UP, 1999: 239-256.

Prince Herndl, Diane. "Reconstructing the Posthuman Feminist Body Twenty Years after Audre Lorde's *Cancer Journals*." *Disability Studies: Enabling the Humanities*. Ed. Sharon L. Snyder. New York: MLA, 2002: 144-156.

Rilke, Rainer Maria. Poems 1906 to 1926. Trans. J.B. Leishman. New York: New Directions, 1957.

Rizk, Mysoon. "Reinventing the Pre-Invented World." The Art of David Wojnarowicz. Ed. Dan Cameron. New York: Rizzoli, 1998: 45-68.

Rosenblum, Barbara and Sandra Butler. *Cancer in Two Voices*. San Francisco: Spinsters Ink Books, 1996.

Savran, David. *Baltimore Waltz*. New York: Theatre Communications Group, 1996: iii-xii.

Scarry, Elaine. *The Body in Pain: The Making and Unmaking of the World.* New York: Oxford UP, 1985.

Schneider, Rebecca. "On Taking the Blind in Hand." *The Body in Performance*. Ed. Patrick Campbell. New York: Routledge, 2001: 23-38.

Scliar, Moacyr. *The Centaur in the Garden*. Transl. Margaret A. Neves. Madison, Wisconsin: U of Wisconsin P, 2003.

Shlain, Leonard. *Art & Physics: Parallel Visions in Space, Time, and Light*. New York: Morrow, 1991.

Shusterman, Richard. Practicing Philosophy: *Pragmatism and the Philosophical Life*. New York: Routledge, 1996.

Siebers, Tobin. *The Body Aesthetic: From Fine Art to Body Modification*. Ann Arbor: University of Michigan Press, 2000.

Silver, Marisa. "Night Train to Frankfurt." *The New Yorker*. 20 (2006): 74-85.

Sobchack, Vivian Carol. *Carnal Thoughts: Embodiment and Moving Image Culture*. Berkeley: University of California Press, 2004.
Sontag, Susan. *AIDS and its Metaphors*. New York: Farrar, Straus, Giroux, 1989.
—. *Illness as Metaphor*. New York: Farrar, Straus and Giroux, 1978.
—. *Regarding the Pain of Others*. New York: Picador, 2004.
Spence, Jo. Cultural Sniping: *The Art of Transgression*. New York: Routledge, 1995.
—. *Putting Myself in the Picture: A Political, Personal, and Photographic Autobiography*. Seattle: The Real Comet Press, 1988.
Storr, Robert. "When This You See Remember Me." *Félix Gonzales-Torres*. Ed. Julie Ault. Göttingen: SteidlDangin, 2006: 5-38.
Takemitsu, Tōru. *Confronting Silence: Selected Writings*. Trans. Yoshiko Kakudo and Glenn Glasow. Berkeley: Fallen Leaf Press, 1995.
Thatcher, Eugene. "Database/Body: Digital Anatomy and the Perception of Medical Simulation." *Images of the Corpse: From the Renaissance to Cyberspace*. Ed. Elizabeth Klaver. Madison: University of Wisconsin Press, 2004: 169-185.
Terry, Jennifer. "The Seductive Power of Science in the Making of Deviant Subjectivity." *Posthuman Bodies*. Eds. Judith Halberstam and Ira Livingston. Bloomington : Indiana University Press, 1995: 135-162.
Turner, Bryan. "Biology, Vulnerability, and Politics. "*Debating Biology: SociologicalReflections on Health, Medicine and Society*. Eds. Lynda Simion William and Bendelow Gillian. New York: Routledge, 2003: 271-282.
Tzu, Lao. *Tao te Ching*. Indianapolis: Hackett Pub. Co,1993.
Vogel, Paula. *Baltimore Waltz*. New York: Theatre Communications Group, 1996.
Vasterling, Veronica. "Butler's Sophisticated Constructivism: A Critical Assessment." *Hypatia*. 14.3 (1999): 17-38.
Waldby, Catherine. *Visible Human Project: Informatic Bodies and Posthuman Medicine*. New York: Routledge, 2000.
Watley, Simon. *Imagine Hope: AIDS and Gay Identity*. New York: Routledge, 2000.
Weinstone, Ann. *Avatar Bodies: A Tantra for Posthumanism*. Minneapolis: University of Minnesota Press, 2004.
Whitefield, Sarah. *Lucio Fontana*. Berkeley: University of California Press, 2000.

Wilcox-Titus, Catherine. "Of Madness, Eroticism and Spirituality." 14 April 2005. <http://www.worcestermag.com/archives/2005/04-1405/current/artsandentertainment.shtml>

Williams, Marjorie. "Hit by Lightning: A Cancer Memoir." *Woman at the Washington Zoo: Writings on Politics, Family, and Fate*. New York: Perseus Publishing, 2005: 307-339.

Williams, Simon J. *Medicine and the Body*. Thousand Oaks, California: SAGE Publications, 2003.

Wojnarowicz, David. *Close to the Knives: A Memoir of Disintegration*. New York: Random House, 1991.

Yang, William. "Allan from Sadness: A Monologue with Slides." *Portraits in the Time of AIDS*. Eds. Thomas W. Sokolowski and Rosalind Solomon. New York: Grey Art Gallery & Study Center, 1988: 34-51.

Young, Katherine. *Presence in the Flesh: The Body in Medicine*. Cambridge: Harvard UP, 1996.

Zaner, Richard M.M. *Conversations on the Edge: Narratives of Ethics and Illness*. Washington, D.C: Georgetown UP, 2004.

About the Author

In the summer of 2007, Catalina Florina Florescu earned her Ph.D. in Comparative Literature from Purdue University. While there, she taught Mythology, Latin, and also served as a researcher within the Comparative Literature Department, working on two generous grants offered in support of Purdue's most aspiring scholars. During the academic year of 2007-2008, she worked at Rutgers University as a lecturer of Expository Writing, and she also taught writing at St. Peter's College during 2009-2010.

She has been recently awarded an MLA's International Bibliography Program fellowship. Dr. Florescu is interested in the multifarious manifestations of the writing process, with a special emphasis on the idioms of pain and suffering, and has published book chapters, essays and articles. Her most rewarding sources of inspiration are the loss of her parents and the joy of Mircea, her son. Her future projects will be creative pieces, as she has started to work on (and publish) plays and short novels.

Contact: http://catalinaflorescu.blogspot.com/